CHICKENS FIRST:

UNTANGLING MYSTERY OF SPECIATION

Jianyi Zhang

JYZ Books, LLC
Clarksville, MD

Printed in the United State of America

Library of Congress Cataloging-in-Publication Date

Jianyi Zhang, 1955-
    Chickens First: Untangling Mystery of Speciation/by Jianyi
    Zhang

ISBN: 0-9743272-0-4
1. Evolution (Biology)   I. Title

Library of Congress Control Numbers (LCCN)
2003095488

Published by:
JYZ Books, Inc.
5816 White Pebble Path
Clarksville, MD 21029
    p. cm
    Includes index bibliography.

To order book, use one of the following methods:

1. http://chickensfirst.net
2. Paypal: enter email address: payment@chickensfirst.net
3. Send $19.95 plus $4 for S/H in money order to:

JYZ Books, Inc.
5816 White Pebble Path
Clarksville, MD 21029

Maryland resident should add $2 for state tax.
International buyer (outside USA) should pay $8 for S/H.

Contents

Table Contents

Figure Contents

Preface

Charles Darwin wanted to provide a firm ground for belief in evolution and persuaded his readers that natural selection was the mechanism to drives evolution. In his first objective, Darwin was successful. Within a few years of publication of the book, the majority of intellectuals were convinced of evolution. However, in his second aim Darwin failed because his theory wrought chaos and factions it created. Darwin's theory of natural selection is still the most influenced theory of evolution and has been subject to considerable debates ever since its inception.

My interest in natural selection was sparked by religious seminars where Darwinism was attacked. The lecturers often cited the lack of intermediate fossils as major scientific drawback. In the ensuing years, I have realized that the Darwin's theory cannot adequately provide the etiology of speciation. What is the mechanism of speciation? Darwin and his contemporaries could not answer the question correctly, because today's scientific knowledge was not available then. Contemporary knowledge of biology, paleontology and genetics has uncovered the inherent weaknesses of the theory. One needs the faith to accept natural selection as the mechanism of speciation as to accept religious doctrine.

This book is not an argument about evolution. Here, I provide a novel mechanism of speciation that is simple, intuitive, and biologically plausible. This model provides many unrelated and unsolved mysteries in biology with reasonable explanations. All of the mechanisms are derived from current biological knowledge.

This book has several divisions. Chapters 1-3 provide readers with a basic background of the biological and genetic aspects of evolution and speciation. For readers with advanced knowledge, you should begin with chapter 4.

Chapter 4 discusses several major hypotheses that have influenced research of the issue. In the chapter 5, a new model of speciation is detailed. In the chapter 5 to 7, I discuss numerous questions on the mechanisms; most of those questions have not been properly explained by any theory of speciation. That Darwinism has had an impact on morality is not disputed, in the chapter 8, I explore moral implications of the new model, while the chapter 9 suggests several avenues for confirming the mechanism. Finally, the chapter 10 summarizes the book.

When the book was nearly finished, I found a lengthy article on Internet, which raises over two dozen challenges against natural selection. These questions are scientifically sound, and might be utilized to argue the proposed mechanism. They are in the appendix I with my comments.

This book differs from many other scholarly works. It is written for readers who have understanding of biology at the level of a high school. More advanced readers should seek out other references.

I express my gratitude to Michael Behe, Jiabei Wang, Jonathan Wells, and Shibin Zhou for their input. The author is grateful to for their helpful comments, suggestions, discussion, and criticism. All remaining errors are mine and mine alone.

Chickens First

Chickens or eggs? Which came first? This ageless question underpins how a new species evolved. The answers from evolutionary biologists are insufficient to appease the inquisitive. Given that shortcoming, I have titled my book with the answer.

Chapter 1

Genetic Primer

One of the most important discoveries of the 20th century was the discovery of the DNA molecule and its structure. It has had a powerful effect on all areas of biological research. In this chapter, I will give a brief overview of genetic principles that underpin life. To understand the molecular genetics of my proposed mechanism of speciation, it is more important to learn the basic concepts than the intricacies of these processes. Due to limited space, I briefly introduce basic concepts that are important for the following chapters. Readers should refer to standard textbooks for detailed explanations.

1-1. DNA, chromosomes, genes and mitochondrial DNA
        1-1-1. DNA
        1-1-2. Chromosomes
        1-1-3. Genes
        1-1-4. Mitochondrial DNA
1-2. DNA replication, RNA, and proteins
1-3. Protein
1-4. Mutations
        1-4-1. Somatic, germline, and zygote mutations
        1-4-2. Causes of mutations
        1-4-3. Types of mutations

1-1. DNA, chromosomes, and genes

Genetics is primarily concerned with the understanding of the ability to store, express, and transmit hereditary information from one generation to the next generation.

1-1-1. DNA

Most living organisms (except for viruses) start as a tiny sphere of cells no larger than a dot on this page. Within that microscopic ball, over six feet of deoxyribonucleic acid or DNA is coiled up. DNA is the genetic material that controls all aspects of development. Genes are the simple units of genetic information encoded into DNA.

In 1953, James Watson and Francis Crick published a paper in the scientific journal *Nature* describing their research that used chemical and physical data from Erwin Chargaff, Rosalind Franklin, and Maurice Wilkins to propose the double-stranded helix structure of DNA, which is composed of two chains (also called strands) that twist to the right to form a helix.

DNA resides in the nucleus of living cells. It is a long polymer made up of nucleotide subunits each containing a phosphate group, the 5-carbon sugar deoxyribose, and a nitrogenous base. The four nucleotide bases are adenine (A), cytosine (C), guanine (G), and thymine (T). Genes consist of the complicated DNA chemical structure. The nucleotide bases of the DNA molecule form complementary pairs: the nucleotides hydrogen bond to another nucleotide base in a strand of DNA opposite to the original. This bonding is specific, and adenine always bonds to thymine (and vice versa) and guanine always bonds to cytosine (and vice versa).

1-1-2. Chromosomes

The chromosome, a nuclear organelle, is the starting point of gene expression. Supercoiled DNA composes its superstructure. Humans have 46 chromosomes, or 23 pairs. One member of each pair is maternally derived, while the other is paternally derived. For the first 22 pairs, each member is identical to the other. The 23rd pair is the sex chromosomes. Human males have one long (X) chromosome and one shorter (Y) chromosome; females have two identical X chromosomes. The condition of having two sets of chromosomes (and therefore two sets of genes) is called *diploidy.*

Chromosomes serve two basic functions. They carry genes that encode protein synthesis. They also can replicate themselves during cell division, thus allowing the cell to create an identical copy.

1-1-3. Genes

Certain portions and sequences of DNA are called genes, which, besides maintaining the body's daily activities, also determine heritable characteristics such as height, skin pigmentation, and eye color. A gene is located at a particular location on a chromosome, called its *genetic locus.* One gene comes from the organism's father and the other from its mother. The combination of the two genes at a locus is called a *genotype.* The two copies of a gene in an individual may be the same, or slightly different (i.e., the amino acid sequences of the proteins encoded by the two copies may be identical or have one or two differences). If they are the same, the genotype is a *homozygous;* if they differ, it is a *heterozygous.* The different forms of the gene that can be present at a locus are called *alleles.*

3

1-1-4. Mitochondrial DNA

In addition to the DNA present inside the nucleus of cells, some DNA lies outside the nucleus in tiny organelles called mitochondria. They are the powerhouses of cells to provide energy for cell metabolism. Mitochondrial DNA (mtDNA) codes for key components of our energy production system, whose purpose is to extract energy from food molecules suspended in the cytoplasm outside the nucleus. Unlike nuclear DNA, recombination does not occur in mtDNA. It is only inherited maternally; therefore, paternal mtDNA is not passed on. Sperms only contain a small number of mitochondria in their tails to propel their movement swim toward the egg, and these mitochondria are cast away with the tail when the sperm head penetrates the egg at fertilization.

Mitochondrial DNA is more susceptible to random errors in copying than nuclear DNA, because it lacks sophisticated proofreading machinery. These mutations also make it easier to resolve differences between closely related individuals and possibly determine evolutionary history.

With these unique properties, mtDNA has opened up new possibilities for tracing evolutionary lineages of living populations. Since the 1980s, many species have been tested with the assumption that complex similarities in mtDNA are attributable to shared matrilineal ancestors.

1-2. DNA replication, RNA, and proteins

A critical property of DNA is that it serves as a template for the synthesis of other macromolecules. In the terms of nucleic acids, the complementary base pairing provides the mechanism, by which information can be transferred across generations. Because each

4

strand of the DNA double helix contains a nucleotide sequence that is the exact complement of the sequence of its partner, both strands contain the same information. Each strand can, therefore, serve as a template for the synthesis of a complementary strand containing the same information.

The actual enzymatic steps involved in the replication of DNA are quite complex, but the overall principles can be stated quite simply. Replication begins with separation of the two complementary DNA strands in a local region. Each strand then serves as a template for a new DNA molecule by the sequential polymerization of nucleotides. Selection of the correct base to be added at each step depends on its being complementary to the next base in the parental template strand. Eventually the replication process generates two complete DNA double helices, each identical in sequence to the original. DNA replication is *semiconservative* because each daughter DNA molecule contains one of the original parental strands plus one newly synthesized strand.

Proteins are not synthesized directly from DNA, but from RNA or ribonucleic acid. Although RNA structure is similar to DNA, their differences are significant. RNA gets its name from the sugar group in the molecule's backbone – ribose. Like DNA, RNA has a sugar-phosphate backbone with nucleotide bases attached to it. RNA contains the bases adenine (A), cytosine (C), and guanine (G), but substitutes uracil (U) for thymine (T). RNA is a single-stranded molecule and is the main genetic material in viruses.

Three types of RNA direct protein synthesis: messenger RNA (mRNA), transfer RNA (tRNA), and ribosomal RNA (rRNA) in both prokaryotes and eukaryotes. The mRNA specifies the amino acid sequences of proteins. The base-pair information that potentially specifies the amino acid sequence of a protein is called the genetic code. Each amino acid is specified by a sequence of three nucleotides called a codon. Other sequences of the DNA specify where the RNA copy is to stop.

After mRNA transcript enters the cytoplasm, tRNA synthesizes polypeptides by providing amino acids to the ribosome that correspond to individual's codons from the mRNA. Ribosomes are the binding sites in the cytoplasm for all molecules involved in protein synthesis. Once the protein chain has been completed, it then undergoes extensive processing that includes folding and linking with other protein chains before it is a mature complex. In addition to the three types of RNA mentioned above, eukaryotes also have soluble RNA that attach the correct amino acid to the protein chain that is being synthesized at the ribosome of the cell according to directions coded in the mRNA.

1-3. Proteins
Cells function through a myriad of biochemical pathways with each step catalyzed by an enzyme or enzymes. In 1941, George Beadle and Edward Tatum, working with mutations of the fungus that imposed nutritional requirements on the organism, demonstrated a firm relationship between genes and enzymes. Enzymes are a special subset of proteins that catalyze chemical reactions without being destroyed.

Since enzymes catalyze the steps, Beadle and Tatum proposed the one gene—one enzyme hypothesis. This hypothesis has later been modified to the *one gene—one polypeptide hypothesis,* since not all proteins are enzymes and not all enzymes consist of only one polypeptide.

Proteins, like nucleic acids, are linear polymers; they are single unbranched chains of amino acid building blocks. An *amino acid* is a small molecule that contains an amino group (NH_2) and a carboxylic acid or carboxy group (COOH) plus a variable side chain. Individual amino acids are linked by peptide bonds, whereby the amino group of one amino acid is joined to the carboxy group of another amino acid. Because proteins are joined by peptide bonds, they are often described as *peptides* or *polypeptides,* although these terms are usually reserved for small proteins.

Polypeptides are constructed from 20 different amino acids. By incorporating so many different amino acids, each with their chemically diverse properties, proteins have greater functional versatility than either DNA or RNA. The specific properties of proteins depend not only on the linear sequence of their amino acid building blocks (primary structure), but also on their folded three-dimensional characteristics (secondary and tertiary structure).

Living organisms are complex systems. Hundreds of thousands of proteins exist to perform all bodily functions. These proteins are produced locally, assembled piece-by-piece to exact specifications. An enormous amount of information is required to manage this complex system correctly. This information, detailing the specific structure of the proteins inside of our bodies, is stored in DNA.

In addition to elucidating the structure of DNA, Crick proposed a genetic principle entitled "The Central Dogma". The dogma states that DNA makes RNA, RNA makes proteins, and the genetic information is from DNA to proteins, not verse versa.

1-4. Mutations

A mutation is a change in the nucleotide sequence of DNA. When a cell reproduces, its DNA and genes are physically replicated. For various reasons, replication can accidentally introduce an error. Such errors are called *mutations*. The DNA repair machinery in the cells may instantaneously correct the error. Moreover, the repair process itself can make a mistake and introduce a mutation. The new sequence of DNA that results from a mutation may code for a protein with different properties that may or may not have deleterious effects on the organisms that carry the mutant genes.

1-4-1. Somatic, germline, and zygote mutations

Somatic cells are body cells such as the bone marrow, liver, or skin cells. Germline cells are the gametes such as sperms and eggs. Fertilized eggs are also known as zygotes. The significance of mutations is profoundly influenced by the distinction among these cells.

When mutations can occur in a somatic cell, the consequence are 1) nothing, 2) damage to cellular function, 3) cancer arises, or, 4) cell death. When a cell with a somatic mutation dies, the mutation no longer exists unless the cell has replicated itself prior to death.

Detection of germline mutations and measurement of mutation rates are very problematic in diploid cells. Most forward mutations (normal gene to mutant form) are recessive; the mutation won't be detected, unless a zygote gets two copies of the mutant on the same gene.

Zygote mutations, in contrast, will be found in every cell descended from it. In other word, if the mutants survive, every cell would contain the mutation. Included among these will be the gametes, which pass the mutation to the next generation.

1-4-2. Causes of mutation
     A mutagen is a natural or human-made agent (physical, chemical or biological), which alters the structure or sequence of DNA. Mutagens can be classified into three groups:

A. Chemical mutagens
     In 1942, Charlotte Auerbach reported the first mutagenic action of a chemical when she showed that nitrogen mustard (component of poisonous mustard gas used in World Wars I and II) could cause mutations in cells. Since then, many other mutagenic chemicals have been identified. For example, bromouracil, an artificial compound extensively used in research, resembles thymine and it will be incorporated into DNA and pair with adenine like thymine.

B. Physical mutagens
     Radiation itself was discovered in the 1890s when Roentgen discovered X-rays in 1895, Becquerel discovered radioactivity in 1896, and Marie and Pierre Curie discovered radioactive elements in 1898. The mutagenic effects of radiation on genes were first reported in the1920s.
     X-rays and gamma rays produce reactive ions (charged atoms or molecules), when they react with biological molecules; thus, they are considered ionizing radiation. Ultraviolet radiation, an important mutagen, is not an ionizing agent; but it can react with DNA and other biological molecules to cause mutation.

C. Biological mutagens

Viruses, plasmids, and transposons are biological agents that can modify DNA structures drastically. Numerous viruses can cause malignant alterations of chromosomes. For instance, Simian virus 40 (SV40) has induced chromosomal changes reflecting a process throughout neoplastic progression of human fibroblasts (Ray, 1995).

The Epstein–Barr virus (EBV) is a human virus that causes an asymptomatic infection. However, it has been linked to the development of several malignant tumors, including B-cell neoplasms such as Burkett's lymphoma and Hodgkin's disease, certain forms of T-cell lymphoma, and some epithelial tumors. All these tumors are characterized by the presence of multiple extrachromosomal copies of the circular viral genome in the tumor cells and the expression of EBV-encoded latent genes, which appear to contribute to the malignant phenotype (Murray, 2001).

Plasmids are molecules of DNA that are found in bacteria separate from the bacterial chromosomes, and are transferred from one cell to another cell in replicative fashion - that is both cells contain the plasmid in the end. Some plasmids integrate into the chromosome and mobilize the chromosome into recipient cells. Large segments of the chromosome may be mobilized from the donor cell to the recipient cell.

Transposons are segments of DNA that can move around to different positions in the genome of a single cell. In the process, they may cause mutations with increase (or decrease) the amount of DNA in the genome. These mobile segments of DNA are sometimes called "jumping genes". As a major cause of spontaneous genetic change, they are now exploited as important tools for isolating genes, and for studying gene expression. If a transposon inserts itself into a functional gene, it will probably damage it. By inserting into exons, introns, and even into DNA flanking the genes (which may contain promoters and enhancers), it can destroy or alter the gene's activity.

1-4-3 Types of mutation

Mutations, or heritable alterations in the genetic material, may be gross or at the level of the chromosome, or point mutations (this technically means mutations not visible as cytological abnormalities and/or those which map to a single "point" in experimental crosses). The latter can involve just a single nucleotide pair in DNA.

A. Point mutations

A triplet of three nucleotides that encodes genetic information is called a codon. When the base sequence of a codon is permanently changed (e.g., GCA is changed to GAA), a point mutation has occurred. There are four types of point mutations: (1) missense, (2) nonsense, (3) silent, and (4) frameshift mutations. A missense mutation occurs when a change in base sequence converts a codon for one amino acid to a codon for a different amino acid. A nonsense mutation occurs when a codon for a specific amino acid is converted to a chain-terminating codon. A silent mutation is a mutation that converts a codon from an amino acid to another codon that specifies the same amino acid; thus, it is unlikely that a silent mutation will affect the behavior of a cell. A frameshift mutation occurs when a nucleotide is deleted or added to the coding portion of a gene, which frequently jumbles a large portion of the information encoded in a gene and can produce aberrant protein products.

Phenylketonuria (PKU) is caused by a point mutation, in which the enzyme that converts the amino acid phenylalanine to tyrosine is defective.

Around 50 years ago, it was discovered that if PKU infants are identified and placed on a special low-phenylalanine diet early in life, the devastating neurological sequelae of excess phenylalanine could be avoided. Due to effective screening and the availability of dietary intervention, thousands of children have been saved from severe mental retardation. Most children can discontinue the low phenylalanine diet by adolescence when the critical period for brain development has passed.

Point mutations are rare events. From in vivo and in vitro studies, the overall human mutation rate is estimated to be about once every million nucleotides per generation. This rate is similar to those measured in various prokaryotic and eukaryotic microorganisms. Humans inherit 3×10^9 base pairs of DNA from each parent. This means that each cell has 6 billion (6×10^9) different base pairs susceptible to mutations.

The overwhelming majority of DNA does not have any effect on the organisms, they are "neutral". They may locate on repetitive regions like Alu elements and other so-called "junk" DNA; or noncoding regions of DNA. Even they are in coding regions, the existence of synonymous codons may result in the altered (mutated) gene with the same amino acid in the protein.

B. Chromosomal mutations

A change in the organization of a chromosome or chromosomes is called a chromosomal mutation or chromosomal aberration. The changes could occur on any part of a chromosome, and it also can affect more than one chromosome at the same time.

1). Structural variations in chromosomes

Chromosomes may deviate from their normal structure in many ways. These aberrations include:

o **Deletion**: an aberration in which a segment of a chromosome is missing. This segment may be large enough that it is microscopic, or so small that only sophisticated methods of molecular testing can detect it. Deletions are detected less often, because they will likely be lethal events.

o **Duplication**: an aberration in which a segment of a chromosome is repeated and thus is present in more than one copy within the chromosome. Duplications are one of the most important mechanisms of genetic rearrangements in both prokaryotes and eukaryotes. They also can be microscopic or submicroscopic.

o **Translocation**: an aberration in which a chromosome segment is transferred to another nonhomologous chromosome.

o **Inversions:** an aberration in which a chromosomal segment has reversed its orientation within the intact chromosome.

2). Variations in Chromosome Number

Polyploidy

Polyploidy is the presence of whole sets of chromosomes in excess of the normal, or euploid, number; the number of chromosomes is a multiple of the normal diploid number. Polyploidy can arise in three ways:

o Errors during gamete formation in which the chromosomes duplicate but the cytoplasm and cell as a whole fail to duplicate and separate into two daughter cells.

o Errors at fertilization, the simultaneous fertilization of a haploid egg by two haploid sperm (resulting in three sets of chromosomes, or *triploidy*).

o Errors in the early mitotic cell divisions of the developing embryo, in which chromosomes duplicate but cytokinesis (separation of the cytoplasm) fails to occur.

Aneuploidy

Aneuploidy is the addition or loss of individual chromosomes from the normal set. Aneuploidy is most often caused by nondisjunction, the failure of chromosomes to separate properly during mitosis or meiosis, most often during the latter.

Aneuploidy in human is often seen as monosomy, the loss of a chromosome, resulting in 45 chromosomes, or trisomy, the gain of an extra copy of a chromosome or chromosomes, resulting in more than 46 chromosomes. The Down's syndrome is the most known chromosomal abnormality in human, with 47 chromosomes, 3-piece chromosome 21.

Contrary to the point mutation, the rate of gross mutation is much higher. It is estimated that 50% of spontaneous miscarriage is associated by chromosomal abnormalities. From studies of human infertility, ten to twenty-five percent of infertile couples carry abnormal karyotypes on their chromosomes (Sankoff, et al., 2002; Kalantari, et al., 2003).

14

Chapter 2

Concepts of Reproductive Biology

Successful reproduction is essential for survival of all species. Reproductive biology is the study of the processes that affect reproduction. This chapter provides readers with an introductory background to understand speciation.

"Like begets like" expresses idea offspring are like their parents. In the strictest definition, only organisms that reproduce asexually meet this criterion. In asexual reproduction, an individual that passes all its genes to its offspring; thus, their genomes are almost 10% copies of their predecessors. All single cell and some multi-cell organisms reproduce asexually.

In sexual reproduction, male and female members of the same species come together to have offspring. Each parent passes on a copy of its genes, so the resulting offspring have two copies of genes from different donors. This chapter focuses only sexual reproduction, a more complicated process that is critical to comprehending the speciation mechanism discussed in later chapters.

2-1. Somatic cells, gametes, and zygotes
2-2. Mitosis
2-3. Meiosis
2-4. Development of an embryo

2-5. Twins and supertwins
    2-5-1. Monozygotic and dizygotic twins
    2-5-2. Supertwins and mixed identical
        supertwins

2-1. Somatic cells, gametes, and zygotes
    The cells that make up tissues or organs (e.g., stomach, intestines, skin, bones, leaves, and petals) are somatic cells. They are a particular cell type and can give rise only to tissues, for which they are programmed. The majority of organisms have a clear male-female distinction, whose gametes are produced by germ cells. In somatic cells, chromosomes are diploid, while gametes are haploid.
    Fertilized eggs, or zygotes, result from the union of a sperm and an egg. They have the potential to become any type of cell - eye, brain, liver, etc. This ability to become any type of cell is known as totipotency. A totipotent cell becomes cells that differentiate into tissue or organs. Any alterations that occur in the DNA of zygotes can be passed onto their descendents.

2-2. Mitosis
    Mitosis is the method by which a cell duplicates itself leading to each daughter cell receiving an identical copy of its genetic material. At the end of mitosis, there will be two identical cells instead of one.
    Prokaryotes and eukaryotes differ in this aspect. Prokaryotes have a single, simple circular DNA molecule in double helical to store genetic information. Eukaryotes, such as humans, have much larger, linear strands of double stranded DNA to form chromosomes.
    Cell division is a continuous process that has four phases: prophase, metaphase, anaphase, and

telophase. The details of each cell phase will not be addressed here. Briefly, during the process, the tightly packaged duplicate chromosomes separate, move to opposite sides of the cell, and pinch off the cell membrane to form the membranes of two cells. The two new cells have identical DNA.

During mitosis, one cell with 2n chromosomes divides to become two cells with 2n chromosomes: the number of chromosomes per cell is conserved; the replication of DNA preceding the division of the cell has prepared chromosomes with two chromatids. The material has been multiplied by two before the division takes place: the quantity of material per cell is conserved; the replication of DNA results in two identical chromatids: the quality is preserved.

2-3. Meiosis

Haploid cells, such as gametes, are formed by a special cell division called meiosis. The double set of chromosomes is reduced to one set. During meiosis, DNA replication is followed by two cell divisions. At the end, one parent cell has produced to four gametes with only one chromosome of each homologous pair. DNA does not replicate during the short interphase between the two cell divisions. In the first division, the number of chromosomes per cell is reduced. The second division conserves the number of chromosomes, but divides the chromatids.

During meiosis, chromatids randomly exchanges bits of themselves, or crossover. This process is called genetic recombination, which results in new genetic structures. An individual who possesses two different forms of a gene or chromosomal recombination is *heterozygous*.

During meiosis, homologous chromosomes line up side by side along their entire length in a process called *synapsis*. The pair of chromosomes, called a *bivalent*, is held together at certain points of their corresponding parts, and some of the matching parts are exchanged between the two chromosomes. If the chromosomes have significant structural alterations, they may not line up correctly. The exchange chromosomal material is unequal leading to one chromosome that has extra material and one chromosome that has missing material. Individuals with chromosomal abnormalities may be infertile or have reduced fertility. Thus, the effects of the abnormalities are very important in reproductive biology.

2-4. Development of an embryo

After fertilization, the resulted zygote is a single cell, which rapidly divides. At the 32-cell stage, the zygote becomes a morula. If the chromosomal complement of the zygote is aberrant, development may not progress beyond a certain stage, even development does progress, the possibility still exists for embryonic or fetal death. A liveborn with a chromosome abnormality can range from phenotypically normal to anomalous.

With additional cell divisions, the morula becomes an outer shell of cells with an attached inner group of cells; the cell mass in the developing embryo is in the blastocyst stage. The outer group of cells will become the membranes that nourish and protect the inner group of cells, which will become the embryo.

After eighth week, the cells of the embryo not only multiply, but also begin to function. Differentiation has produced the various cells and organs that make up a multicellular organism such as a human. Gestation varies across species with humans requiring 40 weeks complete development.

2-5. Twins and supertwins

Twins are two offspring born from the same pregnancy. Supertwins are more than two offspring in the same pregnancy. Both twins and supertwins can be identical or fraternal.

2-5-1. Monozygotic and dizygotic twins

There are two types of twins: monozygotic (identical) and dizygotic (fraternal). Identical twins result from the splitting of a single fertilized egg. These twins are genetically identical in their genotypes and phenotypes. The identical does not mean they are 100 percent the same. Random errors can arise during the processes of DNA replication, production of RNA and proteins after the splitting, which will cause slight differences between the twins; mother can tell one from other, even they are identical twins.

Fraternal twins arise from two fertilized eggs and are no more alike than siblings at different pregnancies from the. same parents. They may or may not be concordant for sex. The probability of having twins is influenced by race, maternal age, and numbers of previous multiple gestations.

2-5-2. Supertwins

"Supertwins" is a common term for triplets and other higher-order multiple births such as quadruplets or quintuplets. These babies can be identical, fraternal, or a combination of both. In human, triplets occur in one in 7,000 births, whereas quintuplets are likely to be born only once in 47 million births. In nonhuman animals, supertwins are commonplace. Insects have hundreds of supertwins at one time.

The supertwins can be identical of fraternal. In the fraternal supertwins, each egg fertilized by a single sperm and developed into baby in utero. If a zygote self-replicates into multiple identical eggs with the same genetic makeup, those zygotes will become identical supertwins with successful births.

If two zygotes in the opposite sex self-replicate, they will become numerous zygotes in different sexes, or the multiple mixed identical zygotes (MMIZ), which would develop into full-term babies. The babies are multitudinous infants in opposite sexes, or mixed identical supertwins (MIST). The process is shown on figure 2.1.

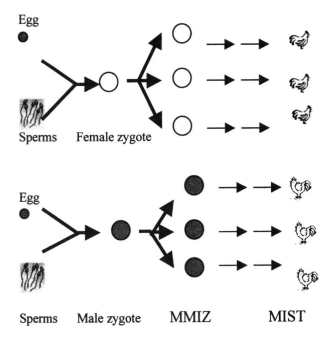

Figure 2.1. Formation of the mixed identical supertwins (MIST) through the multiple mixed identical zygotes (MMIZ).

Chapter 3

Evolution and Speciation

Evolution has been a controversial topic since the publication of Darwin's book. The question is not if evolution ever occurs, but how it occurs. Speciation or generation of new species is the core of the issue. This chapter introduces the basic concept of species and speciation to prepare readers for topics in the subsequent chapters.

3-1. Evolution
3-2. Major scientific methods to study evolution
    3-2-1. Fossil records
    3-2-2. Comparative anatomy
    3-2-3. Molecular biology
    3-2-4. Biogeography
3-3. Species concepts
    3-3-1. Morphological concepts of species
    3-3-2. Biological concepts of species
    3-3-3. Recognition concepts of species
3-4. Reproductive barriers
    3-4-1. Temporal isolation
    3-4-2. Behavioral isolation
    3-4-3. Mechanical isolation
    3-4-4. Gametic isolation
    3-4-5. Postzygotic isolation

3-1. Evolution
    What is the evolution? Evolution means change of characteristics passed to the next generation by hereditary.

    Most nonevolutionists are perplexed by the definitions of biological evolution because even evolutionists do not agree on the definitions. One of the most respected evolutionary biologists has the following definition.

    "In the broadest sense, evolution is merely change, and so is all-pervasive; galaxies, languages, and political systems all evolve. Biological evolution ... is change in the properties of populations of organisms that transcend the lifetime of a single individual. The ontogeny of an individual is not considered evolution; individual organisms do not evolve. The changes in populations that are considered evolutionary are those that are inheritable via the genetic material from one generation to the next. Biological evolution may be slight or substantial; it embraces everything from slight changes in the proportion of different alleles within a population (such as those determining blood types) to the successive alterations that led from the earliest protoorganism to snails, bees, giraffes, and dandelions (Futuyma, 1986)."

    Evolutionary arguments are usually dichotomized into "microevolution" and

24

"macroevolution." Microevolution or change below the species level may be considered as minute changes in the functional and genetic constituencies of populations. These observations are undisputed by critics of evolution. Macroevolution is regarding generation of new species, or speciation, which, however, is vigorously challenged. Micro- and macroevolution have different definitions by different authors. In this book, macroevolution is an initiation of a new species; whereas microevolution is only the change of gene frequencies within a species, and it does not lead to speciation.

Neo-Darwinists maintain that macroevolution is a continuation of microevolution, and accumulation of tiny genetic mutations for a long time would eventually lead to macroevolution.

Abiogenesis is the theory that life can arise spontaneously from non-life molecules under proper conditions, and is not discussed in this book, as it has no direct relationship to speciation. In the ensuing discussions of evolutionary theory, it is assumed there is species pre-existed regardless of their origins.

3-2. Major scientific methods to study evolution

Numerous scientific methods have been used to study evolution. These methods include fossil records, comparative anatomy, molecular biology, and biogeography.

3-2-1. Fossil records

A fossil record is evidence of past life. When an organism dies, its soft tissues are eaten by scavengers or are decayed by microbial action. Body parts such as shells, bones, and teeth are often fossilized. For this reason, the fossil record is biased for species with skeletons. Organisms that consist mainly of soft parts such as worms and plants are unlikely to fossilize. If they were preserved, researchers could have determined much about the original living form.

Additional to biased preservation; fossil study has two other drawbacks. First, a fossil record is subject to conflicting interpretation. Second; the ancestor—descendant relationships cannot be determined through fossil records (Wells, 2000).

To determine fossil age, geologists date past events using relative and absolute techniques. Radioactive dating techniques can infer absolute geological time in number of years; the examined material has a built-in geophysical clock. The most famous is Carbon-14, better known as C^{14}, is used to date organic materials such as shells, bone, or wood. An isotope of carbon, C^{14}, enters the cells of all living organisms from food and accumulates throughout their life. When animals die, their C^{14} begins to change to an isotope of nitrogen (N^{14}). The half-life of this radioactive decay of is 5,720 years. From the proportion of C^{14} remaining, one can estimate when the plant or animal died. The C^{14} method has an effective range of 50,000 years. Beyond that, only a little C^{14} remains, and effective dating is unachievable. Isotopes with longer half-life are available.

Relative dating would tell if an object is older or younger than surrounding objects, and can be inferred by correlating unknown fossils with known ones. The inference of fossil age of surrounding igneous rocks illustrates relative time measurement. If we know the relative date of an object, then we know dates of other objects around it.

3-2-2. Comparative anatomy

Comparative anatomy was first developed when adaptations to special conditions were viewed as deviations from the older form. Georges Cuvier could predict what type of fossil animal was entombed in a rock by examining a protruding bone or tooth.

Comparative anatomists reasoned backwards in time to postulate what ancestors of living forms looked like. A transitional type between birds and reptiles was predicted and, the fossil *Archaeopteryx,* a bird with teeth, claws and a lizard-like tail was found in 1861. however, the method proved unreliable since the history of life is unique and ancestors cannot often be reconstructed from their descendants which may have diverged in wide directions.

Comparative anatomists not only compare existing functional organs, but also vestigial organs and organs called atavisms that are related to a remote evolutionary ancestor. A vestigial organ is an organ that was once useful in an earlier stage of the evolution. Now that organ has become non-functional or underdeveloped, but remains in the body.

An *atavism* is the reappearance of a lost trait specific to a remote evolutionary ancestor and not observed in the recent ancestors of the organism. Atavisms have several essential features: 1) presence in adult life, 2) absent in parents or recent ancestors, and 3) extreme rarity in a population (Hall, 1995).

The results from studies have provided strong evidences for a lineal relationship between an ancestor and the descendant. A close relationship between the organs provides support that a structure was once present in a distant ancestor and has been lost along evolutionary pathways.

3-2-3. Molecular biology

Prior to the 1960s, researchers could only deduce the relationship of organisms by comparing anatomy or physiology. All living organisms from bacteria to humans contain DNA. The sequences of amino acids in an organism's proteins are determined by the genes coding for them. During reproduction, the DNA sequence is replicated; mutations can make the copy slightly different from the parent molecule. Thus, organisms may have DNA and proteins that differ from their ancestors.

In the mid-1960s, Emile Zuckerkandl and Linus Pauling of the California Institute of Technology proposed a revolutionary strategy, which, instead of focusing on anatomy or physiology, analyzed family trees for differences in selected genes or proteins. This approach is known as molecular phylogeny. Much of the early work in molecular phylogeny relied on proteins. With the development of modern techniques of molecular biology, it became more common to analyze the genes coding for proteins than the proteins themselves. In addition to proteins and DNA, all organisms contain RNA, a close chemical relative of DNA that is involved in converting information from DNA into protein sequences. Part of this process relies on tiny particles in the cell called ribosomes, which consist of ribosomal RNA or rRNA. Since 1980, the DNA sequences that code for rRNA have provided much data for molecular phylogeny.

Molecular biologists now compare the "distances" of proteins, DNA, mitochondrial DNA, and RNA seeking to establish evolutionary relationships and the time points of divergence with the biochemical comparative methods.

3-2-4. Biogeography

Biogeography is the science that attempts to document and understand spatial patterns of biodiversity. It studies the distribution of plants and animals over Earth in both a spatial and temporal context, in the past and present, and the related patterns of variation in the numbers and kinds of living things. Three fundamental processes comprise biogeography: evolution, extinction, and dispersal.

The study of evolution by biogeography dates to Linnaeus, Darwin, and Wallace when exploration of the world began and species distribution was observed. Biogeography is not a laboratory science, because the spatial and temporal scales are too large for experimentation, it is a comparative observational science.

3-3. Species concepts

What is species? The term species is derived from a Latin word meaning a kind or class of thing. The ancient philosophical terms *genus* and *species* were used in logic to classify objects and ideas as well as living organisms. The species was a collection of objects that had a common underlying "*essence*". For more than two hundreds years, biologists have struggled to come up with a consensus definition. Biologists recognize three major concepts of species (Mallet, 1995).

3-3-1. Morphological concepts of species
The most distinct feature of species is their appearance. A giraffe, for example, looks like different from a horse. This is called the *morphological concept of species,* which emphasizes assessable anatomical differences between species. Most known species have been designated as separate species based on morphological criteria. Linnaeus in the 18th century adopted this essentialist view of classification in his ambitious project to catalogue every organism known at the time. Linnaeus pioneered the use of *binomial nomenclature*, Latin scientific names for the formal descriptions of species.

3-3-2. Biological concepts of species
In the 1930s and 1940s, evolutionists became dissatisfied with the Linnaean character-based definitions of species. They believed that species classification should represent underlying biological differences than taxonomic convenience. EB Poulton, T. Dobzhansky and E. Mayr proposed that species populations do not interbreed with other species to have reproductive offspring, and are reproductively isolated from other species. The idea is now known as the *biological concept of species (BCS).*

According to BCS, a species is a group of individuals that can only mate and have the offspring among themselves, not with members of another species. A new species of organisms therefore arises when they become sexually incompatible with members from the ancestral species, but breed among themselves with the offspring. Some organisms such as bacteria do not fulfill this definition, because they reproduce asexually. In this framework, it is almost impossible to determine if similar appearing fossils belong to the same species, they might not breed to have the offspring, and also it is difficult to apply it to any organisms that do not exist any more.

3-3-3. Recognition concepts of species

Within a single habitat in the United States, 30 or 40 different species of crickets might breed. Male crickets have songs to attract females. Interbreeding is restricted among species, because each species has its own distinctive song and females will only recognize males that sing their species song.

The *recognition concept of species* emphasizes mating adaptations that become fixed in a population as individuals "recognize" certain characteristics of suitable mates. Paterson has stressed that species can be defined as a set of organisms that share a recognition system (Paterson, 1986). Song, color, and movement all could constitute a recognition system, or specific mate recognition system (SMRS). Defining species by SMRS, rather than reproductive isolation has several advantages. The most important one is that the recognition concept may represent more accurately what happens when a new species originates. However, just like the biological concept of species, it has limitations. For example, scientists may not identify whether a group of animals belonging to the same species, if their recognition mechanism is unknown and/or not recognizable to scientific study, and it is also difficult to apply the concept to any organism not alive any more.

3-4. Reproductive barriers

Although the biological concept of species has limitations, it is the one most appreciated among biologists. It highlights isolation mechanisms that bar genetic transmission among species. Reproductive barriers exist between species, so that they cannot create reproductive offspring. The barriers are divided into two categories: prezygotic and postzygotic barriers.

A number of biological mechanisms prevent interbreeding among different species. Isolating mechanisms preserve the integrity of each species' gene pool, as gene flow between them is impossible. Most species have more than one isolating mechanism.

Prezygotic isolating mechanisms prevent fertilization from taking place. Because male and female gametes never meet, an interspecies zygote does not form. Prezygotic isolating mechanisms include temporal, behavioral, mechanical, and gametic isolation.

3-4-1. Temporal isolation

Genetic exchange between two groups is prevented because they reproduce at different times of the day, season, or year. In fruit flies, type A (*Drosophila pseudoobscura*) and type B (*Drosophila persimilis*) have overlapping ranges, but they do not interbreed. Type A is sexually active only in the afternoon, while type B is active in the morning. Similarly, two species of sage, *Salvia,* overlapping ranges in southern California. Black sage flowers in early spring, whereas white sage blooms in late spring and early summer.

3-4-2. Behavioral isolation

Closely related species may reside in the same geographical area, but they may live and breed in different habitats within that area. Courtship is an exchange of signals between a male and a female,. most animal species have unique courtship behaviors. In general, a male approaches a female and gives a visual, auditory, or chemical signals signal. The female in the same species might recognize the signals and responses to it with hers. The correct exchanges of signals results in mating.

3-4-3. Mechanical isolation

A mechanical isolation refers to morphological or anatomical barrier. Sometimes individuals of different species court and even copulate, but their genital organs are incompatible and successful mating does not occur. The interbreeding of insect species such as dragonflies is thwarted in this manner.

3-4-4. Gametic isolation

Even if mating has occurred between members of two species, their gametes still may not join. Inhibition of mating via molecular and chemical differences between species is called gametic isolation. Fish release their eggs and sperm into the surrounding water, but interspecies fertilization is rare as the egg's surface contains unique proteins that bind only to complementary molecules on sperm cells of the same species.

3-4-5. Postzygotic isolation

When prezygotic isolating mechanisms fail, postzygotic isolating mechanisms arise to prevent successful reproduction. Even mating does occur; the resulting offspring will be infertile.

Embryonic development of an interspecific hybrid is lethal. Development is a complex process requiring the precise interaction and coordination of many genes. The genes from parents of a different species rarely interact properly to have normal embryonic development. In this manner, reproductive isolation is achieved by hybrid inviability. For example, almost all the hybrids die in the embryonic stage when the bullfrog eggs are fertilized artificially with sperm from a leopard frog.

If an interspecies hybrid does reach maturity, it may be able to reproduce. Hybrid animals may display courtship behaviors incompatible with either parental species;, thus, they do not mate. Hybrid infertility occurs secondary to developmental defects of sexual organs or meiosis. If the two parent species have a different number of chromosomes, synapsis that is the pairing of homologous chromosomes occurs during meiosis cannot occur properly. For example, a mule is the hybrid offspring of a female horse ($2n = 64$) and a male donkey ($2n = 62$), which always results in infertile offspring ($32 + 31 = 63$).

Occasionally, an interspecific fertile hybrid develops and produces a second (F_2) generation from a cross between two hybrids or between a hybrid and one of the parent strains. When these hybrids mate, their offspring are anomalous or infertile.

3-5. Speciation

The center of debate among biologists is the mechanism of speciation, which embraces a plurality of theories and hypotheses. According to many popular textbooks, speciation occurs gradually via geographic speciation (allopatric) or competitive speciation (sympatric) or abruptly through mechanisms such as polyploidy (instantaneous).

3-5-1. Instantaneous speciation

Most familiar organisms have two sets of chromosomes, one set from each parent. Such organisms are called diploid. Polyploid individuals have more than two sets. They arise through cytological irregularities during cell division or through the fusion of abnormal gametes. Once formed, they are often sexually isolated from their parent population.

Many plant species have their origins in accidents during cell division that result in extra sets of chromosomes, or polyploldy. An autopolyplold is an individual that has more than two chromosome sets, all derived from a single species. For example, an accident during cell division can double chromosome number from the diploid count $(2n)$ to a tetraploid number $(4n)$. The tetraploid can then fertilize itself (self-pollinate) or mate with other tetraploids. However, the mutants cannot interbreed successfully with diploid plants of the original population.

Speciation by autopolyploidy was first discovered early in 20th century by geneticist Hugo de Vries while he was studying the genetics of the evening primrose, a diploid species with 14 chromosomes. One day de Vries found an unusual variant that had appeared among his plants, and microscopic inspection revealed that it was a tetraploid with 28 chromosomes. The plant was unable to breed with the parental primrose (Campbell, 1996).

3-5-2. Speciation by geographical isolation

By standard textbooks, a catastrophic event in the origin of a species occurs when the gene pool of a population is severed from other populations of the parent species and gene flow is interrupted. With its gene pool isolated, the splinter population can chart its own evolutionary course as changes in allele frequencies caused by selection, genetic drift, and mutations occur undiluted by gene flow from other populations.

Allopatric speciation would occur if the initial block to gene flow were a geographical barrier that physically isolates the population. An effective geographical barrier keeps the allopatric populations apart. The efficacy of the barriers depends on the ability of the organisms to disperse based on the mobility of animals or the dispersibility of spores, pollen, and seeds of plants. Doves and other birds easily cross the Grand Canyon, which is an impassable barrier to small rodents. The same bird species populate both rims of the canyon, but each rim has several unique species of rodents.

Whenever populations become allopatric, it is possible for speciation to occur, as isolated gene pools that generate genetic differences by microevolution. A small isolated population is more likely than a large population to change substantially enough to come a new species.

3-5-3. Sympatric speciation

In sympatric speciation, there is no geographic barrier between ancestral and new species; the new is embedded within ancestral species.

A place where a population lives may contain two (or more) different kinds of resource, for example, two species of food plant. Some individuals may use one plant more effectively, and some the other plant. These specialists are likely to be more successful than individuals who are not as effective in using either plant. Specialists who mate among themselves will be exceptionally successful, because their offspring are likely to inherit their specialization.

A drawback of this explanation is that random mating among the different phenotypes and genetic recombination break up any adaptive combinations of genes faster than they can be selected.

It should be kept in mind that these mechanism discussed above are theories in many standard biology textbooks, so far, the only confirmed mechanism is instantaneous speciation; scientific evidences for other types of speciation are speculative and very debatable.

3-6. Patterns of speciation

Most biologists believe two patterns of speciation exist, anagenesis and cladogenesis.

3-6-1. Anagenesis

Anagenesis is the transformation of a single ancestral species into a single descendant species, it is non-branching evolution, and is involved with the extinction of the older, ancestral species and generation of a new species.

3-6-2. Cladogenesis

In contrast to anagenesis, cladogenesis is the transformation of one ancestral species into a few descendants. The parental species are not extinct, only branching occurs with an increase in the number of species. Cladogenesis is more common than anagenesis, which is probably a special case of cladogenesis. Anagenesis could occur, if the parental population becomes extinct that was coincident to the formation of the progeny species, or the parental species is driven to extinction by the progeny species soon after the latter's genesis (Campbell, 2002).

Chapter 4

Major Theories on Speciation

Many biologists has postulated different mechanism other than the Darwin's theory, before or after Darwin published his book. In this chapter, I will review several major hypotheses on the mechanisms of speciation and introduce briefly main ideas of creationism and intelligent design at the end.

4-1. Natural Selection-Darwinism

The concept of evolution was not completely new, when Charles Darwin stunned the world with his revolutionary views. Europeans in the late 17th century were widely discussing evolution in broad terms. Several thinkers, including Darwin's own grandfather, had proposed evolutionary ideas.

In 1859, Darwin published *The Origin of Species by Means of Natural Selection*, in which he summarized all the evidence supporting the idea that all organisms were descended with modification from a common ancestor. In addition, he proposed that natural selection was the mechanism of speciation.

The theory of natural selection is very simple. All organisms had variations in their characteristics, such as strength, height or speed. Organisms with certain characteristics enjoyed an advantage over their peers who did not possess them. For instance, if a new color provided better camouflage for moths, then moths with the color would live longer and produced more offspring. If the variation were inherited, then the organism's characteristics would change over time. If those changes met certain criteria, a taxonomist would classify it as a new species.

What made Darwin's theory of evolution so different from his predecessors is that he proposed a credible explanation (natural selection) as the principle forces with considerable sustaining evidence drawn from observations of the natural world to support his central theses.

Ernst Mayr, a great 20th century Darwinist, listed five major components of Darwin's ideas (Mayr, 1991a):

(1) *Evolution as such.* This is the theory that the world is neither constant nor recently created nor perpetually

cycling but rather is steadily changing and that organisms are transformed in time.

(2) *Common decent.* This is the theory that every group of organisms descended from a common ancestor and that all groups of organisms, including animals, plants, and microorganisms, ultimately go back to a single origin of life on earth.

(3) *Multiplication of species.* This theory explains the origin of the enormous organic diversity. It postulates that species multiply, either by splitting into daughter species or by "budding," that is, by the establishment of geographically isolated founder populations that evolve into new species.

(4) *Gradualism.* According to this theory, evolutionary change takes place through the gradual change of populations and not by the sudden (saltational) production of new individuals that represent a new type.

(5) *Natural selection.* According to this theory, evolutionary change comes about through the abundant production of genetic variation in every generation. The relatively few individuals who survive, owing to a particularly well-adapted combination of inheritable characters, give rise to the next generation.

Common descent is a general descriptive theory that proposes to explain the origins of living organisms. The theory specifically postulated that all the earth's known biota is genealogically related, similar to how siblings are related to one another. In macroevolutionary processes, one species is transformed into another one.

Ernst Mayr has written:

"... One particular cogent reason why Darwinism cannot be a single monolithic theory is that organic evolution consists of two essentially independent processes, as we have seen: transformation in time, and diversification in ecological and geographical space. The two processes require a minimum of two entirely independent and very different theories." (Mayr, 1991b)

Evidence for the first three components was overwhelming; thus, almost entire scientific community accepted them within a few years after the publication. They are now considered the facts of evolution by a majority of biologists. However, debates continued about how species were generated on the last two components of the theories.

In the Darwin's era, the principles of genetics had not been defined. Scientists did not know how characteristics were passed from parents to offspring. At that time, pangenesis was a popular theory on hereditary. It stated that hereditary information was carried by tiny particles that bud from cells throughout a person's body. These particles or 'gemmules' migrated into the reproductive organs prior to fertilization. Thus, every cell in the body contributed to the constitution of the offspring.

Darwin used pangenesis for the basis of heredity that was hypothesized in his 1868 work *The Variation of Animals and Plants under Domestication*. He believed that it brought 'together a multitude of facts which are at present left disconnected by any efficient cause'. The theory itself was deeply flawed and wrong by the current biological knowledge. Yet, it represented Darwin's attempt to explain such diverse phenomena, such as atavisms, the intermediate nature of hybrids, Lamarckian use and disuse, and limb regeneration. Pangenesis was the foundation of natural selection. Unfortunately, many biologists did not recognize it.

In private letters, Darwin acknowledged that no evidence for his theory existed:

"Long before the reader has arrived at this part of my work, a crowd of difficulties will have occurred to him. Some of them are so serious that to this day I can hardly reflect on them without in some degree becoming staggered (Darwin, 1860)".

"Often a cold shudder has run through me, and I have asked myself whether I may have not devoted myself to a phantasy (Darwin and Darwin, 1887)".

4-2. Use-inheritance –Lamarckism

Jean Baptiste de Lamarck (1744 - 1802) was the first one to initiate a theory to explain how and why evolution occurs. In his *Philosophie Zollogique (1809)*, he introduced idea of vitalism to explain a species' ability to adapt. It is better known as the theory of inheritance and of acquired characteristics. According to this view, certain traits of an organism arise via use or disuse. When a trait was used many times, organs with the trait grew stronger, which made the trait commonplace. On the hand, a trait that was not used caused the corresponding organ to weaken and it was not passed on the organism's offspring.

For example, a giraffe developed a long neck from the need of leaves on treetops. Over time, the giraffe's neck muscles were strengthened to support long necks; hence, its offspring were also endowed with long necks. This example has been compared to a human, regardless of genetic makeup, lifting weights to strengthen his muscles, and then being able to pass on that musculature to his offspring.

Lamarck believed the characteristics acquired by species could be passed to the next generation. The giraffe feeds from the leaves of tall trees. If the environment was favorable for the taller trees, they will grow taller. The environment has placed a need on the giraffe to grow a longer neck to reach the leaves. Subsequently, the giraffe's offspring will be born with the longer necks.

In his own lifetime, Lamarck garnered little support for his theory on the inheritance of acquired characteristics. Only when Darwin made the idea of evolution a possibility were Lamarck's ideas revived. Many biologists accepted evolution, but were uncomfortable with natural selection as the reason of speciation and cast around for alternatives. Because of shortcomings in the Darwin's theory, the inheritance of acquired characteristics was made as a more plausible alternative.

Since then, these notions have been disproved. August Weismann, a German biologist, amputated the tails of 901 young white mice in 19 successive generations. Yet, each new generation of mice had full-length tails. In the final generation, the tails were as long as ones originally measured on the first generations. (Rostand, 1961).

Popularity of Lamarckism declined rapidly with the advent of molecular biology. Modern genetics did not permit the explanation of Lamarckian effects. The discoveries in 1950s that DNA and RNA are principles in genes and heredity restore confidence in the materialistic explanation of evolution. Molecular biologists concluded that hereditary information flows only from the nucleic acid to the protein, never in the reverse direction. This implied that changes in the genetic apparatus could only arise through by errors in replication and variation had to be accidental.

However, Lamarckism is still alive today as the disguised basis of evolutionary biology. In many textbooks, students still read that the ancestors of giraffes kept stretching their necks to reach a higher branch of trees, which was why the necks of giraffes were so long.

4-3. Finalism

All the well-known pre-Darwinian evolutionists, such as Lamarck, Chambers, Spencer, considered evolution a goal-directed process. Many naturalists were also clergymen at that time and liked to find a divine design and purpose in all creatures. As they marveled at the supposedly "perfect design" of organisms for their environments such as the structure of an eagle's wings, they also believed the Creator had devised a plan for every creature. They named the focus of their study from the Greek *telos,* which means "purpose."

Finalism can be traced to Aristotle, who recognized it as one of the causes, indeed the final cause. This belief, which was popular with the natural theology and widespread during the Enlightenment, was called the teleological thinking (finalism). Lamarck's theory of evolution, for instance, postulated a steady rise toward final perfection.

By this idea, evolution progressed from lower to higher, primitive to advanced, simple to complex, imperfect to perfect. The idea postulated the existence of a built-in force, which explained the gradual evolution from the lowest bacteria to orchids, giant trees, butterflies, apes, and humans.

4-4. Speciation by macromutation

Unsatisfied with the Darwinian theory, various mechanisms of speciation have been proposed with the aim to replace it with one that is more plausible. In 1886, Darwin's research associate George Romanes published a paper entitled *Physiological Selection: An Additional Suggestion on the Origin of Species* , in which he proposed a mechanism of evolution: two organisms underwent the same type of variation in germ line cells, and would still be fertile or have "physiological complements" between them. He suggested that germ line cells, like cells of all other organs and tissues, might undergo random variations. Variations, for example, in height or eye color were familiar to everyone.

In 1886, however, no one knew what caused variations, but no one doubted their existence. Romanes emphasized one possible class of variations affecting germline cells would make an organism less fertile compared other members of the species, but it would not directly influence somatic characters. Normally the loss of fertility would be disadvantageous because the organism would not reproduce. However, Romanes argued that if two organisms underwent the *same* type of variation, they would still be fertile between themselves. They would be "physiological complements".

Hugo De Vries, a Dutch botanist, experimented with the evening primrose, *Oerzothera Iamarckiana.* This plant has a peculiar "ring chromosome" that sometimes generates large-scale genetic rearrangements, which produce offspring that cannot cross with the parent plant. This was "instant speciation," without any apparent action of natural selection or any slow accumulation of differences.

Austin Clark (1880-1954) was a staff member of the Smithsonian Institution. As a prominent evolutionist, he authored several books and about 600 scientific articles. Disappointed with no supported evidence for cross-species change, in 1930 he wrote *The New Evolution: Zoogenesis.* He disproved the possibility that major types of plants and animals could have evolved from one another. He proposed the alternate theory of zoogenesis, in which every major type of plant and animal must have evolved—not from one another—but directly from dirt and water (Clark,

1930). The evolutionists in the world were shocked, because Clark was the expert and did not believe in trans-species evolution.

One of the first to recognize the evolutionary significance of chromosomes was the geneticist Richard Goldschmidt. He was a wartime refugee from Nazi Germany and a geneticist at the University of California at Berkeley. He devoted his early career to the genetics of the silkworm moth. Goldschmidt was struck by a pattern of discontinuity in his investigations. The characteristics that separated closely related species of moths were not the same features that Goldschmidt saw *within* the species (Goldschmidt, 1940). He questioned how this could happen, if the evolution of a new species flows smoothly under natural selection on variation from the same species?

After 25 exhausting years of work, Goldschmidt decided that he must establish a different way that cross-species evolution could occur. He announced his new concept: a megaevolution in which one life form suddenly emerged completely out of a different one, he called them "hopeful monsters" (Stanley, 1979).

Similar to macromutations, the virus evolutionary theory was proposed (Andeson, 1970), which suggests the evolution of life as epidemic caused by a virus. It describes that evolution occurred through viral infection and reconstruction of individual genes. According to this theory, life has a new ability, shapes had changed, and existing abilities were lost by embedding the gene of the virus in life. The individual gene evolves or degenerates because the gene of the virus influences the gene array of the individual gene.

4-5. Modern synthesis

During the first part of this century, the incorporation of genetics and population biology into studies of evolution led to a Neo-Darwinian theory of evolution that integrated genetic mutations with Darwin's theory. Natural selection then became a process that altered the frequency of genes in a population. The ideas were formulated from several disciplines, including biogeography, systematics, and population genetics. New discoveries have continued to be introduced to it.

In the original version of the modem synthesis, natural selection was the major cause of evolution at all levels. Populations adapted by natural selection, new species arose when isolated populations diverged as different adaptations evolved, and continued divergence due to natural selection differentiated the higher taxa. Modern synthesis recognizes how genetic drift can cause rapid, nonadaptive evolution. However, the major emphases of the synthesis are gradualism and natural selection.

This point of view held steady for many decades; but more recently, the classic Neo-Darwinian view has been replaced by a new concept, which includes several other mechanisms in addition to natural selection. Current ideas on evolution are usually referred to as the modern synthesis described by Futuyma:

"The major tenets of the evolutionary synthesis, then, were that populations contain genetic variation that arises by random (i.e. not adaptively directed) mutation and recombination; that populations evolve by changes in gene frequency brought about by random genetic drift, gene flow, and especially natural selection; that most adaptive genetic variants have individually slight phenotypic effects so that phenotypic changes are gradual (although some alleles with discrete effects may be advantageous, as in certain color polymorphisms); that diversification comes about by speciation, which normally entails the gradual evolution of reproductive isolation among populations; and that these processes, continued for sufficiently long, give rise to changes of such great magnitude as to warrant the designation of higher taxonomic levels (genera, families, and so forth) (Futuyma, 1986)".

The modern theory of the mechanism of evolution differs from Darwinism in three important respects:

1. It recognizes several genetic mechanisms of evolution in addition to natural selection. Random genetic drift may be as an important factor as natural selection.
2. It recognizes that characteristics are inherited as discrete entities called genes. Variation within a population is due to the presence of multiple alleles of a gene.
3. It postulates that speciation is (usually) due to the gradual accumulation of small genetic changes. This is equivalent to saying that macroevolution is simply many microevolutions.

The advocates of the modern synthesis often reluctantly elaborate with the exact mechanisms of how speciation occurs under their proposal. George Simpson was responsible for incorporating paleontology and macroevolution into the synthesis. Many years later, Simpson wrote:

"This is not to say that whole mystery has been plumbed to its core or even that it ever will be. The ultimate mystery is beyond the reach of scientific investigation, and probably of the human mind (Simpson, 1951)".

Apparently, Dr. Simpson himself was doubtful that modern synthesis solved the puzzles Darwin left unanswered.

The modern synthesis is a theory about how evolution happens at the level of genes, phenotypes, and populations compared with Darwinism that concentrated on organisms at the levels of individuals. To a large extent, modern synthesis has neither brought new mechanisms of evolution to the arena, nor provided satisfactory answers to concerns from general public and biologists, it is just the use of new terms and definitions. It is not surprised that the theory does not end debates, but creates new questions.

Theodosius Dobzhansky, who is a founder of the modern synthesis, has attempted to clarify these issues.

"Let me try to make crystal clear what is established beyond reasonable doubt, and what needs further study, about evolution. Evolution as a process that has always gone on in the history of the earth can be doubted only by those who are ignorant of the evidence or are resistant to evidence, owing to emotional blocks or to plain bigotry By contrast, the mechanisms that bring evolution about certainly need study and clarification (Dobzhansky, 1973)".

The author's uncertainty about these mechanisms is evident.

4-6. Punctuated equilibrium

By the 1970s, an increasing number of paleontologists had become dissatisfied with this approach, because many of the classic examples of gradual change had not withstood the test of modem techniques. If no genuine cases of gradualism existed, then the argument for treating all cases of sudden change as the result of imperfect evidence was undermined. It might be better to reexamine the evidence in a new light, putting aside the traditional Darwinian assumption of gradualism and choosing instead a model of evolution that would allow for the sudden appearance of new forms as indicated by the fossil record.

In 1972, Niles Eldridge and Stephen Gould revived this idea under the name Punctuated Equilibrium (EP). They agreed that transitional fossils were very rare, and they were not as common as the selectionists predicted. A species goes unchanged for a long time before it is replaced without any transition by a new species that resembles a variation of the old one.

They surmised that a group of creatures was separated from the rest of their species; since the group probably lived in a small inhospitable fringe area, they would be under selection pressure. As a small group, they were able to evolve quickly. Later, they spread, and replaced their parent species.

The theory of punctuated equilibria provides paleontologists with an explanation for the patterns found in fossil records. These patterns include the characteristically abrupt appearance of new species, the relative stability of morphology in widespread species, the distribution of transitional fossils when those are found, the apparent differences in morphology between ancestral and daughter species, and the pattern of extinction of species. However, punctuated equilibria have two major negative aspects. The first one is summarized by Dr. John Smith:

"I don't think the proponents of punctuated equilibrium are particularly lucid in telling us just what they think. The factual argument, which most of us probably would now accept is that if you look at fossil records, you do not see continuous gradual change. Instead you see what they call stasis, which is populations changing very little indeed for long periods of time and then changing in really rather sudden transitions" (Campbell, 1996).

The second disadvantage of punctuated equilibria is that it does not provide a coherent biogenetic mechanism to explain how organism at peripheral environments evolves such rapidly.

4-7. Several challenges to the role of natural selection in speciation

Numerous but widely publicized theories of the evolution were discussed above, none of them contains sound biological mechanisms in their hypotheses. From this point, the focus will be on natural selection, as it is still the most influenced theory of speciation among biologists.

Natural selection has been hotly debated since its inception. The lack of linking intermediates has been the most apparent evidence against it. This book explores a few more the challenges to natural selection, with addition to the missing links. Appendix I lists over two dozens more, which interested readers may explore later.

4-7-1. Genetic mechanisms for the alteration of chromosomal structures

Most speciation has observable chromosomal change, either in structure or in number. Let's assume that chromosome number in human ancestors was 48, whereas human has 46 or 23 pair chromosomes. Humans are composed of a few billion cells in hundreds types. The human ancestors would have to have lost one pair of chromosomes from all these cells simultaneously to evolve into humans. From current biological knowledge, it is impossible to think it occurred under the mechanism of natural selection.

4-7-2. Overcoming reproductive barriers

Assuming speciation did occur by natural selection from unknown mechanism, the new species needs to find the same species at the same time in a very close geographical location along the evolutionary pathway across a few million years. If one fly were generated somewhere along the pathway, then it had to mate with another fly in the same species to have offspring. If the second fly were generated a few miles away, the chances of meeting would be dismal. If the second were generated in the following year, then they would have died without reproduction. Had the speciation occurred according to Darwin's theory, it would be mandatory to randomly generate a group of organisms with same genetic structure simultaneous in a very close geographical location. These logistics are impossible.

4-7-3. Which comes first, chicken or egg?

For any plausible theory of speciation, the public and biologists would pose the question, which comes first, chicken, or egg? No proposal addresses the issue seriously.

4-7-4. Co-developments of multiple organs

It is very common for new distinct species do not customarily differ in presence or absence of just a few mutations, but in having different, integrated genetic systems, which may involve differences in dozens, hundreds, or thousands of individual genes. The probability of this happening to form one complex structure, let alone many, is virtually nil.

4-8. After modern synthesis

When the public was informed that evolution by natural selection had been proven, the media declared that the evolutionary theory has strong supporting evidences; the challenges arose among evolutionary scientists themselves. In October 1980, 160 of the world's evolutionary scientists met to challenge the concept of evolution. The verbal explosion was resounding. In the following month, *Newsweek* magazine reported that a large majority of evolutionists agreed that the neo-Darwinian theory or modern synthesis was no longer valid. (Lewin, 1980; Taylor, 1983).

Colin Patterson, a senior paleontologist at the British Museum of Natural History, and is in charge of millions of fossil samples, declared that evolution was "positively anti-knowledge" In addition, he stated, "All my life I had been duped into taking evolution as revealed truth." (Ruse, 1981).

4-9. Nonscientific theories on speciation

According to a recent Gallup poll (2001), Americans are about equally split regarding evolution. About 49% believe in evolution and 45% believe in special creation, while 37% of those who believe in evolution also believe God guides the process.

Phillip Johnson, a professor of law at the University of California at Berkeley, has written several books aimed at providing anti-evolutionary apologetics, including one of the most cited recent anti-evolutionary works, *Darwin on Trial*.

To the naive, Johnson's easy oratory style and impressive credentials make a favorable impression. To those who disagree with creationism as a science, Johnson's book only rekindles the old issue. Impotence of Darwinism, and lack of alternative, sound scientific theory not only make most of public skeptical of the mechanism, but also put whole idea of evolution in a fragile position. Here are some excerpts from Phillip Johnson's remark:

"Keep your eye on the mechanism of evolution; it's the all-important thing. Some Darwinists distinguish between what they call the "fact of evolution" and "Darwin's particular mechanism." The "fact" usually just means that organisms have certain similarities, like the DNA genetic code, and are grouped in patterns (mammals, fish, insects and so on). This pattern of nature is uncontroversial. What's controversial is the cause of the pattern, and particularly whether that cause involves a Creator or only a purposeless material mechanism.

The problem with separating the fact from the mechanism is that a so-called fact of evolution doesn't have much scientific content without a testable mechanism for changing one kind of creature into something entirely different, and especially for building the extremely complex organs that all living things possess. Darwin knew this: it's the first major point he makes in *On the Origin of Species.* The pattern of organisms would provide "unsatisfactory" evidence for evolution, he argued, "until it could be shown how the innumerable species inhabiting this world have been modified, so as to acquire that perfection of structure which most justly excites our admiration (Johnson, 1997)".

4-9-1. Creationism

Prior to the mid 1800s, scientists firmly believed that a Master Designer made all creatures.

Those pioneers who laid the foundations of modern science were creationists.

According to the Bible, humans, animals, and the rest of the universe were created in a six- day period 6,000 years ago. Those who believe this version of our origins are called creationists. Creationists believe not only that each species was specially created, but also that evolution did not and could not have happened.

William Paley (1743-1805), in *Natural Theology*, summarized the viewpoint of the scientists. He argued that the carefully designed structures of organisms clearly point to a Designer. If we see a watch, we know that it had a designer and maker. This is known as the "argument by design." Although we could ignore or ridicule these theories, but our scoffing does not change the situation (Paley, 1802).

Doctoral scientists to conduct research on the topic of the creation-evolution founded Creation Research Society in 1963. In 1972, Henry Morris founded the Institute for Creation Research (ICR) in El Cajon, California. It has become the leading anti-evolution organization in the world since then.

The linking of speciation with the Bible has allowed creationism to be challenged as a disguised form of religion. Many Christian denominations do not believe that the Bible should be taken literally on scientific matters and condemn fundamentalists for their oversimplification of the relationship between God and man. Many religions do not condemn evolution, because they do not believes creatures in the world rest on a single act of creation.

4-9-2. Intelligent design

Disappointed with the fundamentalist interpretation of the Bible by creationism, a group of scientists with different backgrounds launched the Intelligent Design movement.

Following are a few excepts from the Discovery Institute (http://www.discovery.org), a main hub for the movement:

> "The theory of intelligent design holds that certain features of the universe and of living things are best explained by an intelligent cause, not an undirected process such as natural selection."

> Scientists and scholars supportive of intelligent design do not describe themselves as "intelligent design creationists." Indeed, intelligent design scholars do not regard intelligent design theory as a form of creationism. Therefore to employ the term "intelligent design creationism" is inaccurate, inappropriate, and tendentious, especially on the part of scholars and journalists who are striving to be fair.

> Unlike creationism, intelligent design is based on science, not sacred texts. Creationism is focused on defending a literal reading of the Genesis account, usually including the creation of the earth by the Biblical God a few thousand years ago. Unlike creationism, the scientific theory of intelligent design is agnostic regarding the source of design and has no commitment to defending Genesis, the Bible or any other sacred text. Instead, intelligent design theory is an effort to empirically detect whether the "apparent design" in nature observed by biologists is genuine design (the product of an organizing intelligence) or is simply the product of chance and mechanical natural laws.

Two major issues arise from Creationism and the Intelligent Design movement. Firstly, they have established the controversy as a dichotomy between "atheistic evolution" and "biblical-literalist

Christianity". Thus, any criticisms against evolution can automatically construed evidence for their ideas. If the issues posed by creationists were contrary to natural selection, then it should be revised or replaced by another scientific theory.

"Virtually the entire "creation science" literature consists of the books and tracts published by the Creation Science Institute. Most are arguments against evolution, based on the logic that if evolutionary theory has flaws, weaknesses or cannot account for data, then creationism is proved correct. Their arguments assume that there are only two alternatives creationism or Darwinian evolutionism (Milner, 1990)".

Secondly, Creationism and Intelligent Design lack persuasive power, as they do not provide any guide how to further study evolution. Accepting either Creationism or Intelligent Design implies that the question has been answered. The Supernatural created all creatures in the world, and further scientific investigation beyond descriptive study is not possible.

George Gilchrist (Jaxon, 2002), a zoologist, wrote:

"If intelligent design theory is a viable alternative to evolutionary theory, then scientists must be using it to devise tests and to interpret patterns in the data they collect. What sense would there be in presenting an idea as a scientific theory if the idea were not actually used by working scientists? The importance of a scientific theory is not related to its popularity with the general public, but to its utility in directing research and explaining observations within a particular field of study (Kuhn, 1962)".

Frankly speaking, the majority of Intelligent Design advocates are not religious zealots. Most of their arguments against the Darwin's theory are scientifically sound, because the theory does not provide a reasonable explanation for how the new species arise. It is the failure of the theory and the void of alternatives forces them to look for nonscientific explanations.

Chapter 5

Speciation Mechanism: Gross Mutations in Cluster and Mutant Inbreeding

Darwin has been elevated perhaps the greatest of scientists, and his name stands for a theory that has grown far beyond his work. Mark Ridley, a leading naturalist, makes the case for natural selection simply because none of more plausible explanations have been proposed. The doctrine of the natural selection will stand until proven otherwise despite its weaknesses.

Several researchers have proposed a hierarchical theory, in which different mechanisms function at different stages of evolution. Natural selection would be considered the mechanism of the adaptive evolution of a population, but not the driving force of speciation. A new species begins as small populations isolated from their ancestral species by accidents such as chromosomal mutations. The isolated population will evolve rapidly. Natural selection fine-tunes the population to its environment with generation-to-generation changes in the gene pool that become adaptive. The events that lead to speciation and episodes of macroevolution may be more due to contingency than adaptation. A new species will persist long enough to be entered into the fossil record, if they have adapted to their environment.

This hierarchical theory provides a thoughtful insight on speciation; however, nobody has determined how the new species was initiated under this premise. In this chapter, the proposed model explains how speciation in animal occurs with a biologically plausible mechanism.

5-1. Processes of speciation

The proposed mechanism of speciation has four major steps; I use viviparous animals (living young not eggs are produced) as the model:

1. Formation of two fertilized eggs of the opposite sex

Animal development begins when a sperm fertilizes an egg to form a zygote. Two fertilized eggs in the opposite sex are formed when two eggs combines with an X and Y chromosome bearing sperm.

2. Gross mutations on the zygotes

In fertilized eggs, DNA synthesis is very active and these eggs are extremely sensitive to mutagens. Mutations might arise from mutagenic insults or errors in DNA replication. The mutations are random processes, determined by the nature of mutants and the microenvironments of the zygotes. These mutations can have deleterious or nondeleterious effects, and could occur at one locus or multiple loci. In addition, chromosomes can be affected through addition, deletion, and translocations. The outcome of these changes will generate a mutant organism with new genetic structures. The gametes of these newcomers will not match parental haploid cells during the fertilization process.

3. Self-replication of fertilized eggs

The mutant zygotes can self-replicate to form mixed multiple identical zygotes with gross mutations (MMIZWGM), which could develop into mixed identical supertwins with gross mutation (MISTWGM).

4 Mating among the siblings from the same gestation

The majority of the mutants would die during the embryo stage, leaving a very small number to survive as MISTWGM. Of these, even a smaller number would become adults. The characteristics of the novelties are determined by how the mutations occur. The mutations would be not only demonstrated in the somatic cells of offspring with the novel characteristics, but also inherited, and passed into their gamete cells.

As the newborns have not demonstrated their ability to reproduce, I call them pre-species novelties (PSN). Most of the PSN cannot overcome reproductive barriers, as they were born in singletons. Some of them belong to a group, i.e. a member in MISTWGM. Since MISTWGM live together, and have the same anatomical structures, neither pre-zygotic nor post-zygotic reproductive barriers are present among the fraternal siblings. Inbreeding would be natural among these siblings with the reproductive second generation as the outcome. The process is illustrated in the figure 5-1.

5-2. Factors affecting reproduction of MISTWGM

Conceivably, with the process above, many more species would have been formed, if all of the MISTWGM had developed into a noticeable species. However, most MISTWGM disappeared without any trace, as they could not become populated species to be seen in fossils. Four factors affect reproduction of MISTWGM.

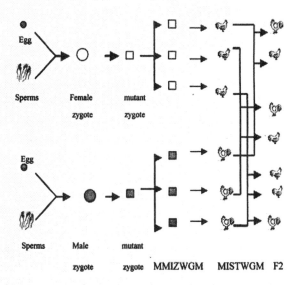

Figure 5. 1. Process of generation of MISTWGM through MMIZWGM.
Note: F2 means the 2nd generation.

5-2-1 Number of siblings from the same birth

According to the mechanism, capability for reproduction of the mutants is directly correlated with the number of siblings in MISTWGM. The more siblings in the same birth, the greater the chance for MISTWGM to overcome reproductive barriers. For example, if 100 MISTWGM with equal number of males and females are born, the proliferation of the next generation by these siblings will be much greater than the situation that only have two siblings born in the same birth.

One challenge to the model is that minimum number is necessary for a species to survive, and only a few members in the "seed" will lead to species extinct at the initial stage. The answer to the challenge is that much more species would have been generated if each MISTWGM had developed into a noticeable species.

According to natural selection, new species evolved from their ancestors gradually that only became possible after long time microevolution. Speciation is a population phenomenon with one or a few branchings during the hundreds thousands, even dozens millions years of the process.

Conversely, by the GMCMI model, generation of MISTWGM could occur at every birth cycle of any multiparious animals. Given almost all members in a species have the potential to give birth to MISTWGM; the number of MISTWGM that had been generated would be astronomical. Majority of these MISTWGM died due to low number of siblings, and many other reasons; however, even only very tiny portion of those MISTWGM become noticeable species, the model would be sufficient to account all of species observed.

Another potential challenge to the model is that many mammals only have singleton offspring and twins are very rare. And how they evolved under the model? It should be remembered that we know how many siblings from the same births occur in their current conditions, but we do not know how many siblings occurred in the ancestral species. While tigers have one or two cubs at a time, their ancestors might have had more in the same litter. Reliable methods are not available to explore how many siblings from the same birth there were in their ancestors.

5-2-2. External, recognizable and stable features

Successful reproduction depends on species members recognizing one another. A species with distinguishable external features or a specific mate recognition system (SMRS) are more likely to meet and mate. The features could be color, odor, body movements, or any apparent external characteristics. Internal attributes would not assist their recognition.

5-2-3. Isolated environments

Isolated environments, per se, do not lead to generation of new species. They do, however, enhance the probability of mating among novelties by keeping the siblings in a close contact. For example, 1000 members of a new fish in a pond without other fish would reproduce more easily than one if they were in an ocean with many other kinds of fishes around.

5-2-4. Coordinated developments of multiple organs

Changes in the genetic structure might make daughter species very different from the parental one with many somatic characteristics; the relevant organs should develop accordingly to accommodate these changes. If a newcomer has much bigger size as a full-term baby, its reproductive system must have corresponding change to adapt the transformation.

5-3. Factors to affect the natural selection of novelties

Generation of PSN is a chance process without any intrinsic principal or teleological force to predetermine the direction. With overcoming reproductive barriers, the new species would proliferate. By the GMCMI model, even a new species was generated randomly, certain features would make them able better to compete with other species, a process know as species selection in evolutionary biology. Three factors determine the selection of whole species.

5-3-1. Enlarged body size

One universal feature is that new life forms typically appear as small organisms that subsequently become larger and more specialized. The horse family with its increasing size over time exemplifies this process. This tendency has also been observed in dinosaurs, titanotheres, and ammonites. (Schindewolf and Reif, 1993).

Increased body size in animals is usually associated with more strength and physical endurance. The characteristics give a new species more survival advantage under the natural selection.

5-3-2. Magic features

A change in the basic design of an organism can produce something magic. Examples of unusual features include wings of insects, flowers of plants, and feathers of birds. If a new species has those features, which their parents did not have, they might be better fit for struggle and survival in competitive world. This is why we see the progress along path of evolution. It is thought that the first birds were from reptiles; the animals with wings would them move to different living environment and they did not need competition with reptiles for food.

5-3-3. Genetic plasticity

According to the GMCMI mechanism, all members of any species would have identical genotypes and phenotypes at the very initial stage. Genetic diversity would develop with the proliferation of MISTWGM, which would allow a population to adapt better to environmental changes. Certain gene complexes have an internal cohesion that is resistant to structure modification. If a critical genetic structure within a species were resistant to modification for some reasons, the species would have high rates of juvenile mortality, problems with captive breeding, and greater risk of decimation by a virus or other disease. Given a new disease or an environment arises, they might be wiped out totally.

5-4. Timing of mutation

The critical role of mutations in speciation has been proposed by numerous biologists. In these proposals no specific timing has been suggested when such mutations occur. In the GMCMI, the mutation arises in the zygotes or dizygotic eggs in the opposite sex. The mutant zygotes become MMIZWGM through self-replication. Each zygote in MMIZWGM cleaves, multiplies, differentiates, and become MISTWGM. The somatic cells of the MISTWGM would have the new phenotype of the mutations; the gametes will have the complete genetic information for the next generation.

Speciation in GMCMI model is a probabilistic process. Theoretically, it could occur under different circumstances. At least, five possible pathways could lead to the production of MISTWGM:

1. The mutation occurs in two eggs and two sperms *separately*, resulting in the same gross mutants. Mutant gametes combine to form two zygotes of the opposite sex, which self-replicate into the MMIZWGM, and eventually become the MISTWGM

2. Two eggs and two sperms combine to forms two zygotes in the *same* maternal body. The mutation occurs on the zygotes separately, resulting in the same gross

mutants. The mutant zygotes become MMIZWGM, and eventually MISTWGM.

3. Multiple eggs and sperms combine to form MMIZ with the mutation occurring *separately*. MMIZWGM arise which eventually become MISTWGM.

4. Two eggs and two sperms combine to forms two zygotes in the opposite sex. The zygotes self-replicate into MMIZ and the mutation occurs in *multiple* zygotes separately resulting in the MMIZWGM that eventually become MISTWGM.

5. Multiple eggs and sperms combine to forms MMIZ that develop into multicellular embryos. In these embryos, the mutation occurs *separately* resulting in the same gross mutants on these embryos. These mutant embryos become MISTWGM.

If we assume each mutation in a single cell occurred independently, four such events happened in scenario 1, two in scenario 2, many more in scenarios 3, 4, and 5. A random event is unique, inadvertent and independent. The requirement of multiple random events with the exact same properties in scenarios 4 and 5 would make them occur against astronomical odds.

Besides the required number of mutation events, DNA replication is not active in the gametes, but very active in the zygotes, which makes the zygotes extremely sensitive to mutagens; thus, these reasons support scenario 2 not scenario 1 as the correct process.

Two mutation events happen in scenario 2, as the two zygotes are very close, the occurrence of the two mutation events is not an independent event any more; thus, the probability to have the exact mutation in the two zygotes would be significantly enhanced. John Davison, a biology professor at University of Vermont, proposed a macroevolution model entitled semi-meiosis (Davison, 1984; Davison, 2000). He suggests that an oocyte instantly acquires a new karyotype in the homozygous state with evolutionary potential as a new diploid organism and "the independent origins of sexual reproduction", which means a different sex in the new species came from an independent process. His model requires that identical mutations occur in the maternal body independently.

A gross mutation is a random process in his model, occurrence of two such mutations with the same properties is very rare; moreover, even if it did occur, they must be in very close proximity and time. If the second event occurs further away, how do they identify each other for successful mating? Requirement of co-occurrence of so many random events makes the model impossible.

If the gross mutation occurs during the later stage of embryonic development, it would require that the exact same mutations occur on entire group of cells. If a change occurs only in part of the embryo, it would probably cause limited somatic defects without affects on gamete structure, the mutant organism would not survive.

The aforementioned model is based on viviparous animals that include most mammals. In oviparous animals, mothers lay eggs and the eggs become liveborns after exclusion. Since the zygotes are grouped together, the supertwins are formed by each individual zygote without re-duplication of itself, the mutations might occur on multiple zygotes in clusters, which is simulated in scenario 3. Table 5.1 is the comparison of these processes.

Table 5.1 Summary of Timing of Mutations

Case	Original germline	Targets of mutations	Number of the events	Probability
1	Two eggs and two sperms	Two eggs and two sperms	4	Low
2	Two eggs and two sperms	Two zygotes in the same maternal body	2	High
3	Multiple eggs and sperms	MMIZ	>>4	Low, but possible in oviparous animals
4	Two eggs and two sperms	MMIZ	>>4	Very low
5	Multiple eggs and sperms	Multicellular embryos	>>4	Very low

5-5. Nature of the mutation

In the GMCMI model, the gross mutation triggers the arrival of a new species, which is defined by its outcome: any modifications of the DNA structure will be the gross mutation, if it leads to failure of bivalent binding between the parental and daughter haploid chromosomes. All chromosomal rearrangements including deletions, duplication, and translocations are considered a gross mutation. Given that over 90 percent of species have changes chromosomal structures, point mutations probably do not have any significant role in speciation.

Mutations can also be non-gross ones. The non-gross ones or micromutations will only modify slightly DNA without generating PSN or MISTWGM, whereas, the gross mutations or macromutations would do so, if the mutants do not die before birth. Both

72

mutations are random processes without a causal relationship between them.

The randomness does not exclude the possibility that certain mutations occur at different frequencies with different mutagens; however, the nature of the mutation does not necessarily have any adaptive value for organisms that carry it. Cold climate may increase the frequency of a mutation in gene X in rabbits, but the mutations neither lead generation of more cold-resistance genes, nor help to generate new cold-resistant animals. The mutant rabbits with more cold-resistance genes would survive longer under cold weather after its birth.

5-6. Stages of speciation

Speciation can be arbitrarily divided into three stages:

1) Birth of PSN

Any gross mutation in the zygotes will create PSN as long as the mutant lives, and it does not matter how many siblings are in the same birth. Only a small percentage of PSN can reach adulthood. PSN would be considered as the first generation of the new species, only if they overcome reproductive barriers and have fertile offspring.

2) Overcoming reproductive barriers by PSN

The PSN who reach adulthood would reproduce themselves, only if PSN are siblings from MISTWGM. Formation of a group of PSN in the form of the GMCMI makes successful mating that produces fertile offspring possible. It is expected that many of PSN have been generated throughout evolution. However, the majority of PSN were singletons, or from small multiple births, they could not reproduce, the next generations were never formed.

3) Natural selection on diverse populations

If the PSN overcome reproductive barriers, the succeeding generations would move around, thus mating would not be limited to siblings from the same birth.

A large population would have dissimilar genotypes and phenotypes to serve as templates for natural selection. Some organisms within the population would be more prolific. Over time, the genotypes of the more prolific members would increase. Natural selection is the only mechanism of adaptive changes; it is defined as differential reproductive success of pre- existing genetic variants in the gene pool. Comparing the number of initial aberrations, only a submicroscopic percentage of PSN developed into a noticeable population to undergo natural selection.

Contrary to Darwin's postulation that new species arise because of natural selection on pre-existing subgroups in population, the GMCMI model suggests that arrival of the novelties is independent of natural selection, and the initial members of any new species have identical genotypes and phenotypes, they are not products of natural selection. Dissimilarity of population would arise unavoidably during reproduction through genetic drift, DNA rearrangement and crossover. Population variations would provide grounds for natural selection. Since almost all existing species, have been assorted populations for a long time; natural selection has eliminated those with disadvantages.

5-7. Answers to the challenges

In the last chapter, I raised several challenges to the theory of natural selection. These scenarios were either biologically implausible or logically impossible. However, with the GMCMI, the puzzles are easily solved.

5-7-1. Modified chromosomal structures

One question that I raise in the chapter 4 is how all members within any species obtain uniform genetic structure in every somatic cell and how the new structures pass to the next generation.

By the model, modification of chromosomal structures can be the outcome of gross mutation that occurs at the zygote stage. The zygotes multiply, proliferate, differentiate into distinct cells with specific functions and the somatic cells demonstrate characteristics of the new species; gametes would have all the genetic information to be passed to the next generation.

5-7-2. Reproductive barriers

How does the new species proliferate and overcome reproductive barriers?

By the GMCMI, any successful speciation should start with a group of identical mutant animals. The siblings cohabitate and have identical chromosomal structures and biological characteristics. Neither prezygotic nor postzygotic reproductive barriers exist to prevent inbreeding.

5-7-3 Chickens first

The chicken or egg paradox has been a puzzle for many centuries. Yet, it has never been the subject of serious scholarly debate.

In chickens, a sperm cell from a male and an egg from female meet and combine to form a zygote - the first cell of a new baby chicken. This first cell divides several dozens times to become the complete animal with billions of cells. Every cell contains almost exactly the same structure of DNA, although expression of DNA of the cells varies greatly.

In the GMCMI model, it is proposed that first group of chickens evolved from a pre-chicken ancestor through gross mutations at the zygote stage. Before the first chicken zygote, only non-chicken ancestors existed. The modified zygotes were the beginning of the first chicken at the one-cell stage. The cells (two zygotes) self-replicated and became MMIZWGM. Then they cleaved, multiplied, differentiated, and eventually became chickens in the form of MISTWGM. Their somatic cells had the characteristics of chickens, while the gonad cells in adult chickens would produce chicken gametes. Therefore, chickens, or a group of chickens, must have come first.

5-7-4. Co-developments of multiple organs

Herbert Spencer pointed out when a new structure evolved, the remainder of the body had to accommodate the new development. Thus, a series of variations would be required to adjust the overall structure correlated to the new organ. What would be the chance for all these variations appearing simultaneously at the right time, if the species had to depend on random variation? Natural selection might explain the changes in a single organ, but not an integrated transformation of the whole body.

The proposed mechanism provides a biologically plausible explanation. Gross mutations might alter many regulatory and structural genes at the same time. Interactions of these genes would make coordinated developments of multiple organs possible in a new organism.

5-8. Direct evidences for the GMCMI model

Several examples directly validate the proposed model. Evolutionary biologists however ignore them. I discuss three examples below and continue to talk about other evidences in the ensuing chapters.

5-8-1. Instantaneous speciation

Instantaneous speciation is a process, in which a new species arise within the range of ancestral populations promptly. Divergence of two gene pools in the same geographical range is a common phenomenon in plants. In 1886, the Dutch botanist Hugo de Vries found unusual dwarf, broad-leaved varieties of the evening primrose growing near Amsterdam. During the next few years, he raised nearly 54,000 plants and named eight new species. They were not formed by the accumulation of slight variations as described by Darwin, but it seemed they appeared suddenly. This sudden speciation was the result polyploidy, which was discussed in the chapter 3.

In another example, Digby (Digby, 1912) crossed the primrose species *Primula verticillata* and *P. floribunda* to produce a sterile hybrid. Polyploidization occurred in a few of these plants to produce fertile offspring. The new species was named *P. kewensis*. Newton and Pellew (Newton, 1929) noted that spontaneous hybrids of *P. verticillata and P. floribunda* produce tetraploid seeds on at least three occasions.

Speciation through polyploidy and hybridization has long been considered less important in animals. A number of reviews now suggest this may be not true. Bullini and Nasceti review (1990) suggests that speciation through hybridization may occur in a number of insect species including walking sticks, grasshoppers blackflies, and beetles.

Although fewer cases of instantaneous speciation in animals have been directly reported, evolutionists have not appreciated the real implication. Instantaneous speciation in both plants and animals are excellent proofs to validate the GMCMI mechanism, which holds that all species arise promptly in the same location as their ancestor. Allopatric speciation, which will be further discussed in chapter 6, did not occur.

Darwinists acknowledge prompt speciation in plants and other organisms, but fail to recognize the mechanism only valid one for any speciation. If instantaneous speciation were only true in some plants, two discrete mechanisms of speciation would exist. However, scientific evidence for evolution supports a single, uniform mechanism.

The reason for lack of appreciation of the instantaneous speciation in animal's world is that the process is much more complicated in animals. A new species starts at one central location as the same site of their ancestors. As nature of animals, they move around because of needs of looking for food, escaping from predators, all of these would make almost impossible to detect speciation in animals by a simple observation.

5-8-2. Purebred animals

The cheetah is one of the most amazing animals in the cat family. As the world's fastest animal, it has been clocked at 110 kilometers per hour for short distances. In 1900, estimated 100,000 cheetahs were estimate worldwide and had, fallen to 30,000 by 1975. In 1997, only 9,000 – 12,000 cheetahs remained in Africa.

Blood samples taken from 50 cheetahs for genetic testing revealed they were genetically identical to each other. Electrophoretic studies have shown that cheetahs are monomorphic and homozygous at many loci, thereby lacking the 10-60% polymorphisms found in other species (O'Brien, 1986). Furthermore, skin graft experiments in cheetahs indicated a significant lack of variability at the major histocompatibility complex (MHC) (O'Brien and Yuhki, 1999). While this compatibility would aid organ transplantation by reducing rejection, the lack of genetic diversity is harmful to cheetahs.

In another example, the pocket gopher lives in tunnels in the American west. Researchers at the University of California, Santa Cruz found out each Humboldt gopher accepted grafts of small skin patches from other members within its own species, whereas the Carmel Valley gophers did not. To test immune function of Humboldt gophers, the researchers grafted skin from Carmel Valley gophers onto Humboldt gophers that rejected the grafts. This result suggests a uniformity of the Humboldt gopher genome (Irion, 1996).

How is the uniformity in both examples achieved? According to bottleneck theory, sudden reductions in population size can alter the resulting gene pools. In the recent past, with change in environmental condition, many individuals in these animals were killed and only a small number have survived. With the drastic reduction in their population, close relatives were forced to breed, and the cheetah became genetically inbred, meaning all cheetahs are closely related. Oddly, no explanation is available to elucidate why and how such kind events only selectively kill cheetahs and leave every other big cats alive to develop its expected genetic variation.

The GMCMI model provides an alternative explanation. Both cheetah and Humboldt gopher might be relatively new species. They have proliferated and become sizable populations recently; however, their genetic structures at MHC are resistant to modification and lacks plasticity. Genetic variation is essential to the long-term adaptation and persistence of populations by providing sufficient genetic options on which natural selection can. Environmental changes over the past 100 years and lack of flexibility in the cheetahs has pushed it close to extinction.

5-8-3. Transplant animals

Transgenic organisms are plants and animals with incorporated foreign DNA in their genomes. A number of approaches are being used to insert foreign genes into plant or animal cells. One approach of generating animal proteins is to use live animals that have incorporated a foreign gene to make the recombinant protein. These transgenic animals are usually produced by microinjecting the DNA of a particular gene into the nucleus of a recipient fertilized egg cell. The eggs are then implanted into the uterus and allowed to develop.

Following a strict definition, transplant animals do not meet the definition of the biological concept of species, by which all species should be natural products. Transgenic organisms were unheard of when the biological concept of species was introduced during 1930s. Nevertheless, in many aspects, the certain transplant animals can be considered as a novel species, they cannot breed with their ancestral species with reproduction of fertile offspring, but could do that with other members from the same experiment. If the genetic structure is stable, and they have external, distinct features, and they have potential to go wild, and become a new species.

Scientists replicate the speciation process in laboratories as Nature has done in the wild.

5-9. Composite of major scientific hypotheses

The evolution mechanism remains a controversial topic. Many biologists have proposed different speciation mechanisms, but have failed to define the complete process.

5-9-1 Macromutations or "hopeful monsters"'

Richard Goldschmidt, a well-known biology professor at University of California, Berkeley, along with the paleontologist Otto Schindewolf, were "saltationists," who view evolution proceeding by sudden "jumps." Goldschmidt thought that large-scale mutations must exist to cause the differences between species. These large-scale mutations could abruptly produce major new evolutionary groups (Stanley, 1979). He coined the phrase—"hopeful monsters"— capturing the sense of his idea of rare, large-scale mutations that would not be lethal, but instead, instantaneously create a new, successful evolutionary group. In his words:

> Subspecies are actually, therefore, neither incipient species nor models for the origin of species. They are more or less diversified blind alleys within the species. The decisive step in evolution, the first step toward macroevolution, the step from one species to another, requires another evolutionary method than the sheer accumulation of micromutations (Goldschmidt, 1940).

Goldschmidt failed to define a plausible biogenetic pathway. While his premise was correct, he could not explain how it happened.

The GMCMI model is compatible with the idea of "hopeful monsters". The newly formed were generated by the gross mutations, not by the accumulation of micromutations, and resulting organisms resemble "hopeful monsters". Additionally, the model proposes that these monsters came to earth in a cluster or group comprised of both genders.

Dr. Ernst Mayr critiqued the idea of "hopeful monsters" (Mayr, 2001), these critiques would get proper answers with the GMCMI mechanism.

Below are some excerpts from his book. I have included my arguments in parenthesis.

"Many different observations and arguments led to the final refutation of transmutationism. First was the realization that a species is not a type that can mutate to a new type, but rather comprises many populations. Not all the individuals in a population can have the same mutation simultaneously. Therefore a new species could not originate instantaneously. Those who postulated that transmutation happens through the origin of a single newly mutated individual were up against other formidable difficulties (*By the GMCMI model, a group of new species was generated instantaneously in the same birth from a single mother. All the individuals in a population had identical genotypes, which only became diverse in later development*). The genotype of an individual is a harmonious, well-balanced system, brought together through millions of years and fine-tuned by natural selection in every generation. Since it was known that potential mutations at most gene loci have deleterious or lethal effects, how could a massive shake-up of an entire genotype by a major mutation possibly produce a viable individual (*By the GMCMI model, even the majority of the gross mutation have deleterious or lethal effects; a tiny portion would survive. A massive shake-up of an entire genotype by a gross mutation possibly produces a viable individuals, which is shown in study of human infertility, further discussed in the chapter 7).* Only an incredibly rare individual (called a "hopeful monster" by Goldschmidt) would have any chance for survival and success, whereas the great mass of such macromutants would be failures. But where are all of these millions of failures resulting from such a macromutational process? *(The vast majority of hopeful monsters does not have SMRS and might not be different from their parents morphologically. Reliable and easy scientific methods are needed to distinguish them from their parents)".*

5-9-2. Specific Mate Recognition System

Hugh Paterson, a geneticist from South Africa, is a well-known field entomologist. He criticized Dobzhansky's ideas of isolating mechanisms in the role of speciation. These mechanisms ranged from simple geographic barriers and disjointed distributions (organisms do not meet to mate) to inherent biological properties that make breeding impossible.

Paterson believed that the factors to facilitate successful mating—constitute a single integrated system: specific mate recognition system (SMRS) that is all necessary to get two separate species—two distinct reproductive communities. Divergence in the two mate recognition systems would yield two species.

With GMCMI, several factors have been suggested to affect successful speciation after the birth of MISTWGM. Reproduction depends their ability to recognize among themselves. A species with more distinguishable external features are more likely to pair successfully.

Several authors such as Authur Cain have proposed a pluralistic species concept (Cain, 1954). Brent Mishler and others have favored a pluralistic species concept that would explicitly state that no single concept accounts for all species. Mark Ridley also presents a similar idea (Ridley, 1996). In the GMCMI model, both biological and recognized concepts of species are emphasized. The hallmark is the biological concept of species, by which one species differs from others internally; however, without SMRS, the newly formed would die out without successful recognization. Both factors are important and necessary the speciation process. Alone, none of them is not sufficient.

5-9-3. Peas bred from a pure line

Wilhelm Johannsen, a Danish biologist, performed very notable experiments that produced ' pure lines' of the self-fertilizing princess bean, *Phaseolus vulgaris*. Studying the progeny of self-fertilizing plants, he selected the trait of bean weight. In his beans, variation showed a continuous range between two extremes. Yet, when this range was analyzed, the continuous range of variation was composed of a number of overlapping but distinct elements. These elements did not blend, but maintained their position in the range over many generations. Based on this, Johannsen asserted that selection is effective only to the extent that it can act on those pure lines lying at the extremes of the variation. Johannsen believed that the only way in which new traits could be introduced into the genome was by a sudden mutation alerting the characteristics of an existing pure line (Johannsen, 1911).

Johannsen's work came at time when genetics was still in its infancy. During inbreeding of pure lines, crossover, genetic recombination, and genetic drift would cause the genetic structures in pure lines beans to become diversified. The different weights of beans were reflections of varied DNA constitutes.

In the GMCMI, all species were pure lines initially; their genetic structures were identical, dissimilarity of the genome would have developed rapidly in species through genetic drift, DNA recombination and crossover in the reproductive process, which also drives natural selection in the later phase.

5-9-4. Natural selection

In this book, natural selection, modern synthesis, Darwinism and neo- Darwinism are interchangeable terms, as they all express that idea that the macroevolution or speciation is an extrapolation of microevolution.

In the GMCMI, the natural selection is given its proper place: it is a major driving force to determine direction of species; individuals with more favorable traits within one species do have chance to have more progenies than their counterparts over long run; however, the offspring would never become a new species just because of it, no matter how long it takes, how isolated they are from other members of same species, the natural selection has nothing to do with initiation of new species or macro-evolution.

Natural selection would keep individuals and species that best fit the environment. These individuals are not biologically superior or inferior; they are just appropriate for their environments. If the environments changed, they might not fit it any more. Species we observe today have been filtered by natural selection for many, many generations. Hence, it is not surprising that they are adaptive to their environment.

Nature also has impacts on species at other levels. For instance, it may favor one species versus others, which is called species selection. It might remove a whole species, just as it deletes alleles in the gene pool. The disfavored species might be wiped out during the evolutionary process.

Natural could favor one group verse others within the same species, which is known as group selection. With group selection, gene frequencies within entire population might not be modified. The confusion is that by Darwinists the group and species selections are not natural selections, as they may not modify genotype frequencies of species. This discussion will continue in Chapter 8.

Chapter 6

Old Stories with New Thoughts

Since Darwin, the theory of natural selection has been attacked from many fronts. Many evolutionists have devoted their careers to find out evidences to defense it. Dr. Jonathan Wells, in *Icons of Evolution,* discussed many textbook cases that support the natural selection. Generally speaking, his criticisms are reasonably correct and all of the well-known cases used by biologists to defense natural selection are either misrepresented or explained with alternative mechanisms, none of those cases provides exclusive evidence for Darwin's theory.

In this chapter, I will discuss a series of cases, which have generated interest among biologists and the public. However, I do not intend to debate the selectionists' explanations. In most cases, I briefly present selectionists' popular views and interpret them under the perspective of the GMCMI model. Readers can ponder these questions and draw their own conclusions.

6-1.   Lack of transitional links
6-2.   Vestigial organs
6-3.   Atavisms
6-4.   Innovative organs
6-5.   Gradualism

6-1. Lack of transitional links

Numerous problems arise from assuming that natural selection is the mechanism for speciation. Darwin himself anticipated the counterarguments and prepared his defense. The most troubling aspect was the absence of a transitional species in the fossil record. Gradual accumulation of differences in evolution by natural selection must include many intermediate forms. Darwin realized his data did not support his theory. He wrote:

'The number of intermediate varieties which have formerly existed on earth must be truly enormous. Why then is not every geological formation and every stratum full of such intermediate links?..(this)…is the most obvious and gravest objection which can be urged against my theory.'

Darwin believed that the intermediate fossils were incomplete. However, after extensive search for more than 140 years, the definitive transitional species are yet found in the fossils.

Creationists attack this vulnerability relentlessly, and the selectionists periodically propose new evidence. Unfortunately, almost all cases of alleged "missing links" have been either frauds or misrepresentations (Wells, 2000a).

The GMCMI model provides multiple explanations for the issue. Firstly, the novelties generated by the gross mutation usually produce huge changes in both genotypes and phenotypes. Speciation through this process, rather than the accumulative changes by microevolution, make the existence of an intermediate with only slight changes unlikely.

Secondly, if speciation was the result of the gross mutations at the individual level, which could occur in each birth cycle of multiparous mothers, and the speed of evolution would be much, much faster than one proposed by natural selection, the quantity of intermediates would be too small to be detected in fossils.

Thirdly, an intermediate with only minor changes could occur, but the changes would not be a part of SMRS. The reason is as follows: to overcome reproductive barriers, MISTWGM need to recognize each other for mating. If a MISTWGM has only one recognizable marker and it differs slightly from the parental species, the newcomers would have little chance to find each other for mating. Assume that the length of the giraffe's neck is the only feature for recognition. If the mean of length in the first generation of giraffes were only slight longer than one in its ancestor, such animals could not have identified themselves among many members of ancestral species; mating between the giraffes MISTWGM and members of ancestral species would not generate fertile offspring.

In 1987, Peter Sheldon published in *Nature* an exhaustive study of approximately 15,000 fossil trilobites from a 3-million-year period (Sheldon, 1987). He studied eight lineages of the small invertebrate marine organisms and focused on the number of ribs in the exoskeleton. Each lineage exhibited a gradual increase in the number of ribs during the 3 million years. There was no evidence to support a long period of equilibrium followed by a brief period of speciation. The significance of ribs in trilobites is unknown. One suggestion is that each rib covered an appendage and provided extra strength. The existence of the intermediate forms, if true, can be understood under the model that the number of ribs was not an important recognizable marker for trilobites, they might have had other markers for recognition, slight changes in numbers of the ribs would not affect adversely their matching.

6-2. Vestigial organs

Comparative anatomy demonstrates that many organisms contain vestigial organs that are organs or parts of organs without apparent function, they often undersized, or lack an essential part.

In humans, the appendix, coccyx, third molars, and muscles that move the ears are remnant organs. Whales and pythons have vestigial hind leg bones, whereas wingless birds have vestigial wing bones. Many blind, burrowing, or cave-dwelling animals have vestigial eyes.

Darwin was interested in vestigial organs, because they conflicted with the view of creation. He wondered how organisms that were the product of a "perfect creation" could have nonpurposeful parts. Biologists always use vestigial structures as evidence for common ancestors in evolution. During evolutionary history, functions have been gained and lost; hence, many organisms should display vestigial structures. This interpretation is reminiscent of Lamarckism use-inheritance.

When an organ loses much or all of its function, it no longer has any selective advantage. As the presence of the vestigial organ is not harmful, selective pressure to eliminate it is weak.

By the GMCMI model, gross mutations lead to new organisms. Innumerable genes might be involved in the process. Structural and regulatory genes responsible for certain organs might be affected with poor development of organ, so that they become vestigial organs.

6-3. Atavisms

Anatomical atavisms are closely related to vestigial structures, they provide another strong evidence for common descent according to Darwinism. An atavism is a structure that was once found in a remote ancestor and now lost in a recent lineage. Mutant horses occasionally displayed gills, this would be considered a potential atavism, because gills are diagnostic of taxa (e.g. fish), to which horses do not belong. For developmental reasons, the occasional occurrence of atavisms is expected under common descent, if structures or functions were lost between ancestor and descendant lineages.

Atavisms also occur at molecular levels. Humans do not have the capability to synthesize vitamin C, which, if deficient in the diet, will lead to scurvy. However, the predicted ancestors of humans could synthesize vitamin C, as do most other animals except primates and guinea pigs. Therefore, humans, other primates, and guinea pigs should carry evidence of this lost function as a molecular vestigial characteristics.

Recently, the L-gulano-g-lactone oxidase gene, the gene required for vitamin C synthesis was found in humans and guinea pigs. It exists as a pseudogene (Nishikimi, Kawai et al., 1992; Nishikimi, Fukuyama et al., 1994). Since it has reported, the vitamin C pseudogene has been found in other primates. We now have the DNA sequences for this gene in chimpanzees, orangutans, and macaques (Ohta and Nishikimi, 1999). Darwinists rarely provide any plausible explanation for how atavisms occur under natural selection.

By the proposed model, gross mutations might affect both structural and regulatory genes responsible for certain organs or chemical products. If only regulatory genes that controlled growth were affected, the organs would not develop in the new animals, the structural genes of these organs would be still intact. The regulatory genes could be reactivated in the descendant lineages, lost organs would reappear. It is conceivable that the structural genes for other vitamins still exist in humans. They might be pseudogenes and/or regulatory genes without normal function.

6-4. Innovative organs

A change in the basic design of an organism can produce innovative or magic organs, a evolutionary mystery in many biology textbooks. Wings of insects and feathers of birds are this group. Usually, these "new" structures are innovative organs.

How do such novel changes occur? The most common explanation by Darwinists is that these organs have values of pre-aptations (formerly called preadaptations); they evolved to fulfill one role, and also had enough evolutionary plasticity to be modified for another. Bird feathers are a good example of a pre-aptation, as they evolved from reptilian scales; the mammalian middle ear is a pre-aptation that evolved from a modified jaw element of reptiles.

The model provides an alternative explanation. As discussed before, gross mutations might alter regulatory and structural genes. Even slight genetic changes could cause major structural changes, if these changes are in right positions, and the results of these changes produce any magic organs, morphological changes generated by the gross mutations are beyond human imagination.

Animals with innovative organs would enjoy wider living spaces and might move into a different ecological niche. Birds with feathers could fly farther and feed themselves easier than their flightless ancestors. A new species with innovative organs is more likely to have survival advantage than the ancestor. The advent of magic organs would lead to progress of evolution: so mutant fishes went to land from sea, the mutant reptiles became birds, which flied into sky.

6-5. Gradualism

By natural selection, speciation should occur in a geological gradual manner. Even gradualism is not a formal mechanism of evolution; it significantly restricts possible macroevolutionary events.

By GMCMI, members of a new species were generated punctually by the gross mutations at the individual level; whereas the natural selection was a gradual process, which became functional only after the novelties became a diverse population. The selection would make every living organism adapt to natural environments over a long time. The initiation of a new species is the instantaneous process; the shaping of species is a gradual one.

6-6. Progress and trend

Evolution has a direction. Since the beginning of life on Earth 3,500 million years ago and the first prokaryotes (bacteria), organisms have become far more diversified and complex. A tiger, a chimpanzee, and an elephant are definitely quite different from a bacterium. Overall, these organisms become more complex. The gradual change over time from bacteria to unicellular eukaryotes, and finally to higher plants and animals, has often been referred to as progressive evolution.

For some Darwinians, the concept of evolutionary progress raises uneasy questions. How can a rigorously opportunistic competitive struggle lead to progress? Darwin himself quietly expressed his doubts. "Never use the words higher or lower," was written on the margin of his copy of *Vestiges*.

The word "progress" can assume different meanings. A teleological thinker would argue that progress is due to a built-in drive or striving to perfection. Darwinians rejected such causation as no genetic mechanism has been identified to control such a drive. However, one can also define progress empirically as the achievement of something that is somehow better, more efficient, and more successful than what preceded it. Progress is indicated by greater complexity, better utilization of the resources of the environment, and all-around adaptation.

Lamarck and other 19th century held that the perfect organism and all forms of life were arranged in a single column based on their assumed progress toward maturity. Now we know that progressive changes in life are neither predictable nor goal-directed.

Progress can be understood under the GMCMI model with explanation similar to magic organs in the section 6-4. More "advanced" animals usually started with organisms that have innovative organs, such organism would obtain new ecological niche, be more competitive and better fit for struggle in nature.

Let's examine the horse family as a classic example for trend of evolution. The fossils of horse show the following characteristics of evolution as summarized by Hunt (Hunt, 1995):

1. Evolution does not occur in a straight line toward a goal, like a ladder; rather, evolution is like a branching bush, with no predetermined goal.
.....
2. There are no truly consistent "trends".
.....
3. New species can arise through several different evolutionary mechanisms.

The evolutionary pattern of the horse is consistent with the GMCMI model. The initial "seed" of any species is the outcome of a random gross mutation. Species generated by the mechanism would be haphazard and highly diverse without any direction. The bush-like pattern was reflection of such randomness. The horses evolved into stronger and faster animals, as the smaller species of horses died out by the effects of natural selection. The trend of increasing size is the outcome of selection on the species level.

6-7. Industrial melanism

Industrial melanism is one of the best examples in textbook to support natural selection, and was considered "the most striking evolutionary change ever actually witnessed in any organism".

The British peppered moth (*Biston betularia*) has two forms: the predominant light form (*typica*) and the uncommon dark form (*carbonaria*). With increasing number of coal-burning factories in England during the Industrial Revolution, it was discerned that the number of dark moths was rising. The moth derived its name from the scattered dark markings on its wings and body. In 1849, a coal-black mutant was found near Manchester, England. Within a few decades, the black form had increased to 90% of the population in this region. This change in color became known as industrial melanism.

In 1896, British biologist J. W. Tutt proposed that the peppered moths might be increased due to camouflage. The moth is nocturnal and rests during the day on tree trunks. In areas without industrial activity, the tree trunks are encrusted with lichen. Light moths are almost invisible against this background. In areas where air pollution is severe, the combination of toxic gases and soot has killed the lichens and blackened the tree trunks. Light moths stand out sharply against such background, and become easy prey for birds. In polluted woods, the dark form has a much better chance of remaining undetected.

In the ensuing decades, many efforts have been to confirm the idea. Dr. Jonathan Wells has put forth sound scientific arguments (Wells, 2000a) to discredit the claim. If the camouflage-predation explanation were true, it would demonstrate that natural selection temporarily altered the proportions of light and dark peppered moths. It does not prove how a new species of moth originates or how new orders or classes evolve. Thus, industrial melanism does not support the role of natural selection as driving force in speciation.

Assuming that the peppered and typical moths are two distinct species, the GMCMI model would provide a proper elucidation regarding how the ratio changed. It can be thought that industrial pollution did not create the peppered moth, in fact, which was generated by a random process under the GMCMI model. Pollution might provide camouflage for the peppered moths, thus making them less subject to predation, and altering the moths ratio.

6-8. Organs and environments

Fishes living in underwater caves are blind with apparent scars where their eyes should be. The eyes of the Mexican tetra have a lens, sclera with a degenerate retina and optic nerve, although the tetra cannot use to see (Jeffery, 2001). Blind salamanders have eyes with retinas and lenses, yet the eyelids grow over the eyes to seal them from outside light (Durand, Keller et al., 1993; Kos, Bulog et al., 2001). Other than the lack of sight, these cave dwelling fishes are identical to fishes living at the water's surface and sighted.

By natural selection, it is thought that a group of fish with normal eyes were trapped swam in an underground cave. Without light stimulation, their eyes atrophied and over the generations, the fishes become blind. However, even in the darkness, the genetic codes in trapped fished still had built-in instructions to develop eyes, these genes did not know that eyes were no longer required, they would keep developing the eyes anyway.

The GMCMI model provides a more plausible explanation. Gross mutations occurred in the ancestral fishes with production of eyeless fish. Some fished went to the surface and did not survive, because they could neither find food, nor escape predators. Some of them swam into lightless caves; the disadvantage of having no eyes did not exist in the caves. Fishes with normal eyes were at a disadvantage. Eyes would be subject to injury from banging into rocks and cave walls in the darkness. Only eyeless fish that remained in the caves flourished.

Fishes without functional eyes are not the outcome of adaptation to an environment. Darwinists consider the environment as the cause and the organ as the consequence; by the GMCMI model, the opposite is true.

6-9. Anagenesis and cladogenesis

Anagenesis is the transformation of one entire species into another one, which is also called phyletic evolution. Cladogenesis, or branching evolution is the budding of one or more new species from a parental species that continues to exist. Most biologists hold that both mechanisms exist. Cladogenesis is more important than anagenesis, not only because it appears more common, but also because only cladogenesis can promote biological diversity by increasing the number of species.

In the GMCMI model, no true anagenesis or phyletic evolution ever occurred in any speciation process. The "seed" of new speciation is generated at the individual level, while the ancestral species remains intact. Evidence pointing to anagenesis is deceiving for two reasons. One, the ancestor was becoming extinct when splitting off the new species happened for unknown reasons. Two, the ancestral fossils after splitting have not been found. In either situation, faculty anagenesis would occur.

6-10. Geographical isolation

An important component of the modern synthesis was Ernst Mayr's idea of geographical speciation, also known as allopatric speciation (Mayr, 1942). Geological barriers could divide a population into two or more isolated populations. For example, a mountain range may rise and split a population of organisms that can inhabit only lowlands. The Isthmus of Panama, a land bridge, may form and separate the marine life on either side. Alternatively, a number of inhabitants may become geographically isolated when individuals from the parent population colonized a new location.

In the GMCMI model, geographical isolation, per se, would never cause speciation. All new species would evolve by the GMCMI mechanism. Geographical isolation plays three roles in evolution. Firstly, members of a new species have to recognize each other for mating. If they were isolated from the parental species, recognition would be easier. Secondly, isolation would prevent individuals from moving between the populations, thus physically separating them. Lastly, isolation could make a small group extinct.

6-11. Drug resistant bacteria

A drug resistant bacterium has been a widely cited case of macroevolution through microevolution in numerous books. Ernst Mayr has written:

"Let us consider drug resistance of bacteria. When penicillin was first introduced in the 1940s, it was amazingly effective against many types of bacteria. Any infection, let us say by streptococci or spirochetes, was almost immediately cured. However, bacteria are genetically variable and the most susceptible ones succumbed most rapidly. A few that had acquired by mutation genes that had made them more resistant survived longer and a few still had survived when the treatment stopped. In this manner, the frequency of somewhat resistant strains gradually increased in human populations. At the same time, new mutations and gene transfers occurred that provided even greater resistance. This process of inadvertent selection for greater resistance continued, even though ever-stronger dosages of penicillin were applied and the period of treatment was prolonged. Finally, some totally resistant strains evolved. Thus by gradual evolution an almost completely susceptible species of bacteria had evolved into a totally resistant one (Mayr, 2001)."

Dr. Wells comments on the drug resistant bacterium in his book *Icons of Evolution*. (Wells, 2000b):

"Most cases of medically significant antibiotic resistance are not due to mutations, but to complex enzymes that inactivate the poison, and which bacteria either inherit or acquire from other organisms. Some cases of resistance, however, are due to spontaneous mutations that alter the bacteria's molecules just enough so an antibiotic can no longer poison them. Bacteria lucky enough to have such mutations (like those lucky enough to have inactivating enzymes) can resist an antibiotic and survive to reproduce."

First, bacteria cannot be categorized as a new species, because they reproduce asexually. Bacteria are often considered a 'strain' in microbiology, rather than a distinct species. A strain is a subgroup within the same species. Many subgroups or strains that are generated by random genetic mutations are found within one species of bacteria. The strains have different characteristics. One strain happens to be resistant to penicillin that offers the strain a growth advantage over penicillin-sensitive bacteria under the drug.

Second, even we assume that drug resistant bacteria is a distinct species, its arrival might not be associated with natural selection or penicillin at all. The drug-resistant organism might have existed long time before the advent of antibiotics through random gross mutations or plasmid infection. It could reproduce along with other bacteria. The use of antibiotics killed other bacteria, the drug-resistant strain easily proliferated and was singled out.

6-12. Bottleneck effects

In biology textbooks, the bottleneck effect or genetic bottlenecks are considered as the result of environmental fluctuations. The most cited case is the cheetah that was discussed in chapter 5.

The proposed GMCMI provides an alternative explanation: in stead of mysterious events which only selectively killed a few types of animals, some of the animals might be new-evolved species, the plasticity of genetic structures in the animals are very poor, they are still identical and very close to the initial stage, when they were created by Nature.

The fossils of cheetah is estimated to have three millions years, the fossils might be cheetah-like animals, or ancestor of the cheetah; they are not the real fossil of the cheetah.

6-13. Adaptations

The premise that living things are adapted for life on earth was accepted before Darwin published the book. John Ray and William Paley, natural theologians, revised it for 18th and 19th centuries. Paley (Paley, 1802) explained that adaptation was the creative action of God, when HE miraculously created everything in the world. The adaptive nature of living creatures is evidence of the greatness of God. Natural theology was a way of understanding adaptations in nature, but its main influence was beyond biology, as it makes a delicate argument to substantiate that God exists. One reason why Darwin's theory was heavily attacked was that it reduces the evidence for the existence of God. The key difference between natural selection and natural theology is that the former explains adaptation on a natural base, which was only explainable by divine action by the counterpart.

The GMCMI mechanism is consistent with Darwinists in this aspect. Natural selection itself is sufficient to provide the explanation for adaptation. However, there is one difference that by most biologists, adaptation is a population-based concept, new species are the products of natural selection, and the members of a new species had been adaptive to the environment at its arrival. In the GMCMI model, the new arrivals are a small identical group; natural selection and adaptation are irrelevant at the initial stages. Natural selection begins to function only the initial "seed" overcome reproductive barriers and the population becomes diverse.

Chapter 7

Questions and Answers

While the questions raised in Chapter 6 have some sorts of answers by naturalists, the ones discussed in this chapter have few, if any. These questions have been grouped in a single chapter here, because they are more challenging and worthy of exploration.

7-1. Human evolution
    7-1-1. Origin of *Homo sapiens*
    7-1-2. Are we alone?
    7-1-3. The future of humans
7-2. Cambrian exploration
7-3. Mosaic of evolution
7-4. Rates of evolutionary change
7-5. Human infertility

7-1. Human evolution

One of the strongest motivations for the development of evolutionary theory is that we want to understand our own past, present, and future as *Homo sapiens*.

7-1-1. Origin of *Homo sapiens*

Where did we come from? This question has haunted human beings for centuries. Our ancestors sought to answer the question with myths. Today, we seek to answer this mystery with science.

Based on fossils discovered in the last 200 years, the two popular theories on human origins are available, the multiregional model and the monoregional model. The multiregional model posits that *Homo erectus* marked the beginning of our global evolution and all subsequent fossils represent evolving *Homo sapiens*. No single place was the birthplace of modern humans. Populations in Europe, China, Java, and Africa all evolved independently.

The monoregional model holds that only one region, Africa, was the site of the complete evolutionary sequence from *H. erectus/ergaster* to modern humans. The evolutionary lineages that developed outside Africa did not give rise to modern humans; they were simply replaced by humans who migrated from their African homeland during the last 100,000 to 200,000 years. This model differs from the multiregional model in that different characteristics of modern regional populations developed during the last 100,000 years or later. Some anthropologists now believe that hominids migrated from Africa to Europe 600,000 years ago or more and evolved into a separate species *Homo neanderthalis*. When *Homo sapiens* spread out from Africa, they did not interbreed with the Neanderthals. *Homo neanderthalis* and all other hominids originating from Africa prior to *Homo sapiens* became extinct for unknown reasons. Only African *Homo sapiens* survived and gave rise to modern humans.

Charles Darwin was an early advocate of the African origin of *Homo sapiens*. He reasoned that humans' closest living relatives the gorillas and chimpanzees are found in Africa perhaps still in the forests where they evolved. As no early hominid fossils have been found anywhere else, Darwin proposed that Africa was the locale to search for them (Darwin, 1871).

The GMCMI model aligns with the monoregional model that all ancestors of *Homo sapiens* were born at single location from a single mother (its non-human ancestor). The probability for such complicated processes occurring at multiple locations and different times does not exist. Since the current discovery of fossils show that all early hominid fossils originated from Africa, East Africa is the most likely cradle of the humans.

7-1-2. Are we unique?

Are there other humans in this vast universe? According to the GMCMI model, the answer is no. Humans and all other species were generated by an inadvertent event that happened after thousands other accidental events. The chance to duplicate these events under different conditions does not exist.

Even an intelligent animal parallel to human intelligence exist somewhere in the universe, the likelihood that they are the exact species just like us is zero. Human is alone and unique in this sense.

7-1-3. The future of humans

Another question lingering is the future of humans. Will humans evolve into another species?

According to the proposed model, it is clear that humans are not intermediate species to any future species. As the overwhelming majority of births in human are singletons despite existence in multiples gestations, the probability for generation of MISTWGM by human is near zero, without even considering inbreeding among siblings if they were born.

7-2. Cambrian exploration

The fossils records show that primitive life on Earth began about 4 billion years ago. In the 3.5 billion or so years that followed, the pre-Cambrian period, those early, single-celled organisms changed slowly. Then, at the beginning of the Cambrian period 545 million years ago, all the major anatomical structures we see today evolved in a span of only 5 to 10 million years in an event known as the Cambrian explosion. That divergence of life led to most of the major animal groups we have known so far.

The Cambrian explosion has been one of the most intensively studied phenomena in biology. Yet, many questions remain about what drove the dramatic differentiation of life early in the Cambrian and why that intense period of adaptation happened when it did. The explanation of this sudden arrival is a scientific conundrum.

The Cambrian explosion challenges Darwinian evolution, because it happened so quickly and so many major groups of animals debuted, it was a remarkable event. The fossil record of Cambrian explosion is definitely not what one would expect of Darwin's theory.

Two issues arise from Cambrian exploration. One, where were the "seeds" of these organisms from? In addition, how was the pattern of exploration formed?

The possibility that an extraterrestrial civilization might have originally "seeded" the earth with the first forms of life was proposed by von Däniken's thesis and supported by Francis Crick (Crick, 1981). Since his pioneering role in elucidating the structure of DNA, Crick has worked on the problem of the origin of life. He argues that the complexity of life is so great, that chances of even the structures arising naturally on the primitive earth are too remote.

A popular belief among astronomers is the existence of planets throughout the galaxy suitable for life (Shklovskii and Sagan, 1966). Some of these planets are older and more suitable for the origins of life than the Earth. A chance exists for intelligent life in the galaxy with a technological civilization that evolved in the past. This civilization would have tried to throw out the easily preserved spores of primitive organisms to set off the process of evolution on many planets. Some indirect evidence supports the hypothesis such as the levels of iridium, rarely found on earth but common in meteors, are elevated in rocks from that period. Many scientists have acknowledged the hypothesis.

The GMCMI model does not offer any clues of the "seeds" regarding how they were generated, as this is not relevant to any speciation theory. The GMCMI model assumes that 'seeds' or spores of organism were already present. However, the pattern of the growth predicted by the model is consistent with what we observed in the Cambrian explosion.

Under proper conditions, the original seeds became primitive organisms, which multiplied and become a viable species that would be the parents for the next new species. The circle would continue in an explosive pattern as shown in the initial phase of the Cambrian explosion. Nevertheless, the trend could not last forever, as ecological niches decreased, food supplies were depleted, and an isolated area was missing. All of these might be contributed to the decreased phase of exploration, as phyla were lost and not replaced.

7-3. Mosaic of evolution

Rates of evolution, speciation, and extinction may differ by several orders of magnitude in different organisms. This uneven rate of evolution of different properties of an organism is called *mosaic evolution*. For example, in *Archaeopteryx*, numerous structures such as teeth and its tail are still reptilian, but its feathers, wings, eyes, and brains are birdlike. Mosaic evolution is even more strikingly demonstrated by the highly different rates of evolution of different proteins and other molecules. The mosaic of evolution was puzzling to Darwin and contemporary top naturalists (Mayr, 1991).

The proposed model encounters no problems to explain the issue. Speciation is the result of gross mutations, in which alterations of DNA fragments might occur in localized loci; majority of genome may not be involved. Any organs controlled by genome with the gross mutation would be altered significantly, whereas organs controlled by non-affected genome would be unaltered.

7-4. Rates of evolutionary change

Rates of speciation have demonstrated great variations in evolution. Certain species of animals and plants have not apparently changed in more than 100 million years. This group includes the horseshoe crab, the fairy shrimp, the lampshell, and plants such as Gingko, *Araucaria*, and *Cycas*.

A well-known illustration of this phenomenon is the evolution of the lungfishes (Westoll, Andrews et al., 1977). The major anatomical reconstruction of this class of fish took place in a span of 75 million years, while almost no further changes occurred in the ensuing 250 million years. Such a drastic difference between the rates of evolutionary change in young and mature higher taxa is the rule (Fig. 7-1). Bats originated from an insectivore ancestor within a few million years, but have not changed their basic body structure in the ensuing 40 million years. The origin of whales happened very rapidly compared to the subsequent essential stasis of the new structural type.

Ernst Mayr believes that the rate of evolution is the only problem without a satisfactory answer (Mayr, 2001).

"In particular, there is one problem that is not yet entirely solved. When we look at what happens to the genotype during evolutionary change, particularly relating to such extreme phenomena as highly rapid evolution and complete stasis, we must admit that we do not fully understand them."

"The complete standstill or stasis of an evolutionary lineage for scores, if not hundreds, of millions of years is very puzzling. How can it be explained? In the case of a living fossil, all the species with which it had been associated 100 or 200 million years ago had either changed drastically since that time or had become extinct. Why did this one species continue to prosper without any changes in its phenotype? Some geneticists thought they had the answer by ascribing it to normalizing selection, which culls all deviations from the optimal genotype. However, normalizing selection is equally active in rapidly evolving lineages. To explain why the underlying basic genotype was so successful in living fossils and other slowly evolving lineages requires a better understanding of development than is so far available."

According to the GMCMI model, every species started as a pure line with a similar genotype and phenotype. They had not undergone the process of natural selection. Through reproduction, a mixed population would develop. Many of the resulting characteristics would not fit the environment, thus the individuals would be removed by nature. Selection eliminated all the individuals of a population who deviated from the optimal phenotype. After the hundreds or thousands of generations of *normalizing or stabilizing* selection, the population would be close to the optimal genotype. The stabilized population would have incorporated the necessary modifications in their phenotypes in many generations, and the resulted animals would not change significantly afterwards.

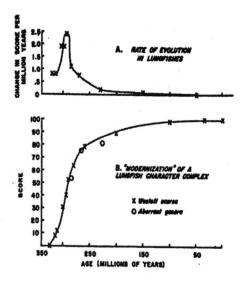

Figure 7.1

Rate of acquisition of new characteristics after the origin of lungfish. (A) Acquisition of new characteristics per million years. (B) Rate of approach to the final kingfish body plan per million years. Most of the reconstruction of the body plan of the new taxon takes place in the first 20 percent of its life. *Source:* Simpson, George G. (1953). *The Major Features of Evolution,* Columbia Biological Series No. 17, Columbia University Press: NY.

7-5. Human infertility

Human infertility is a current medical issue that seems unrelated to evolutionary theory. It has been touched by biologists in their discussions of evolution. I believe it will contribute to our understanding of the speciation mechanism.

Infertility is the failure of a couple to become pregnant after one year of regular, unprotected intercourse. About 8% to 10% of couples in reproductive age experience infertility, and in approximately 40% of these cases, male infertility is the major factor. Another 40% of infertility problems are caused by female factors while the remaining 20% involve both members. The vast majority of chromosomal aberrations do not lead to a live birth. However, a small number of them do survive to adulthood.

Infertility has several known etiologies. One of them is chromosomal abnormality, which accounts for 10-25% infertility in human; there might be much more at submicroscopic levels, which are not easily studied with microscopic techniques. Current world population is over 6 billion. If only 1 percent of the population have a chromosomal abnormality, then there are at least a few dozen million in the world with hundred different kinds of abnormal karyotypes.

Individuals with a chromosomal alteration may appear normal and even have a normal life span. However, by the biological concept of species, these individuals are not the same species as other members of humans, because they cannot breed with them and have fertile offspring. Reasons for their infertility might be due to facts that they have no recognizable markers for their partner even there are individuals with the same genetic structure, as almost all of these individuals do not differ from each other and their normal peers. They might lack reproductive ability as well. Without a proper recognized system or SMRS, even if the reproductive systems function well, they still cannot find the partner with the same genetic structure or in the same species, and they can only live in the world as pre-species novelties (PSN).

From these studies, we have learned that chromosomal aberrations are not necessarily lethal, and a normal appearance and life span are possible. Organisms with the aberrant chromosomal structures are infertile, because they cannot find corresponding counterparts. As PSN, they cannot develop into a new species. They could be reproductive, if they can be matched with individuals who carry the same chromosomal structures, even the chance to have the same structure is very remote.

One of the most common misconceptions among biologists is that macromutation is always lethal and deleterious; the studies on human infertility neutralize this view.

Chapter 8

Survival of the Most Diversified

Since the birth of Darwin's theory, the impact of evolutionary thought on our modern culture and society has been tremendous. To quote Richard Dawkins, "I think that Darwin's theory of evolution by natural selection is the most powerful idea ever to occur to the human mind." (Dawkins, 1998).

"Consider these examples: Marx and Keynes in economics and social studies; Dewey in modern education; Fosdick and 'higher' Biblical critics in modern theology; Nietzsche, James, and Positivists in modern philosophy; Beard in American history; Frankfurter in modern law; London and Shaw in novels; Camus, Sartre, and Heidegger in existential thought; White in sociology; Simpson and Dobzhansky in paleontology and modern genetics; Huxley and P. Teilhard de Chardin in humanism."

"The concept of evolution was soon extended into other than biological fields. Inorganic subjects such as the life-histories of stars and the formation of chemical elements on the one hand, and on the other hand subjects like linguistics, social anthropology, and comparative law and religion, began to be studied from an evolutionary angle, until today we are enabled to see evolution as a universal, all-pervading process (Newman, 1995)."

The erroneous theory and its blind acceptance by intellectuals have caused the society pay huge price with the loss of millions of lives. Discussion of the impact is beyond the scope of this book.

The model not only provides answers to numerous biological issues, but also opens a door for debating moral issues from another stance. In this chapter, I will discuss the social and moral implications of the mechanism.

All natural selection in this chapter refers modifications of variations with the same species.

8-1. How does nature select?
     8-1-1. Selection by traits
     8-1-2. Selection by group
     8-1-3. Selection by chance
8-2. Who survives?
8-3. Selfish and altruistic genes
8-4. The survival of species-the most diversified
8-5. Morale and its biological origin

8-1. How does the nature select?

In chapter 5, I discussed the role of natural selection and suggested that it does not play a notable role in generating a species; however, it is important to shape species afterwards. The question is how natural selection does work.

8-1-1. Selection by traits

Biological variations are within every species, which give each individual different traits or characteristics. Environments will favor certain individuals with particular traits over others, and these individuals will have more offspring.

Selection by traits includes sexual selection and kin selection. An example of sexual selection is observed in peacocks. Male peacocks have large, brilliantly colored tails sexually preferred by females.

Male peacocks fan out their feathers and strut as a mating show for females who select a mate based on color and physical prowess.

Kin selection refers to the altruistic acts that individuals perform for relatives without any derived benefits for themselves. Altruism will evolve if alleles for it are passed from generation to generation more often than ones with nonaltruistic behavior.

8-1-2. Selection by group

Animals are often clustered into different groups. In lower organisms, they are grouped together because of locations. Members of such groups cooperate by watching for predators, sharing food and water resources. This cooperative behavior enhances the survival propensity of such a group.

Human are a highly organized species often polarized and separated by geography, race, politics, and religion. Survival has been highly dependent on which group one belongs to. War might decimate a race or country completely, but biological differences between the victor and the vanquished are insignificant.

8-1-3. Selection by chance

The world is plagues with potentially catastrophic events like flooding, fire, and earthquakes. These events are unpredictable in their magnitude and damage. According to naturalists, individual phenotypes are units of natural selection. Although other types of selection occurred, they are not part of natural selection (Mayr, 2001b),

One might have impression that particular individuals are selected over other members of their species for their traits, such as strength, health, and speed. This premise operates in a very limited range.

Selection functions on multiple levels. Only selection of traits will alter phenotypic frequency, if natural condition worked consistently, and phenotypes are hereditary; selection by group and chance would not be likely to modify the frequency of phenotypes significantly.

8-2. Who survives?

In Darwin's theory, survival of the fittest and struggle for existence are two key phrases that have been challenged philosophically.

The claim that Darwinism is little more than a vacuous tautology has become popular among the philosophical critics of the theory (Manser, 1965; Macbeth, 1971; Bethell, 1976). Their argument starts from the description of natural selection resulting from the "survival of the fittest"—a phrase coined by Herbert Spencer and subsequently adopted by Darwin. The problem is: how do we know which organisms are the fittest? According to critics, the biologists' only answer is that the fittest are those who do, in fact, survive longest. The "survival of the fittest" thus turns out to mean no more than the "survival of those who survive." Natural selection is reduced to a tautology, a principle that is true not because it is confirmed by the facts but because its components are defined in a way that makes its truth a logical necessity. It contains no more useful information than, say, the statement "All husbands are men," which is necessarily true since "husband" is defined as the male partner. The whole concept of natural selection amounts to nothing more than play on words, and as such it cannot possibly constitute a workable mechanism of evolution (Bowler, 1989).

Darwinians naturally view this argument as a misunderstanding of how the theory works (Ruse, 1982): They claim that natural selection is based on the belief that the fittest do survive longer and reproduce more frequently, but fitness is defined not in terms of survival but as a measure of the organism's ability to cope with its environment.

As the selections work on multiple levels discussed above, the survival of the fittest is valid only if every member of the species has equal opportunity to compete within the same species, which rarely occur in the real world. The survivor might not have any better traits than ones did not, as individuals may survive just by chance. Survival of the fittest is also a statistical concept that can only apply to a population. In other words, on average, the ones who survive might be better fit their environments than ones who did not. For an individual, that might not be true. One cannot apply population statistical concepts to each individual in the species.

For those who better fit their environment, this is just a transient state. Environments change often unpredictably, and ones that fit environment may become unfit instantly under new environments.

8-3. Selfish and altruistic genes

English evolutionist Richard Dawkins's popular book, *The Selfish Gene*, emphasizes the genetic aspects of evolution. He was not as much concerned with the biological basis of behavior as he was with the biological basis of selfishness and altruism.

Since all organisms inherit all their genes from their ancestors, rather than from their ancestors' unsuccessful contemporaries, all organisms tend to possess successful genes. They have what it takes to become ancestors—and that means to survive and reproduce. This is why organisms tend to inherit gene with a propensity to build a well-designed machine—a body that actively works as if it is striving to become an ancestor. That is why birds are so good at flying, fish so good at swimming, monkeys so good at climbing, viruses so good at spreading. That is why we love life and love sex and children: it is because we all, without a single exception, inherit all our genes from an unbroken line of successful ancestors. (Dawkins, 1976).

Are there genes that make humans or any organisms successful? As the title suggests, Dawkins argued that genes are selfish: they will do whatever it takes to ensure that their carrier—the individual—replicates these genes. Evolution, therefore, has ensured that our behavior brings about the preferential survival of the genes we carry. Those behaviors are "selfish", because they preserve our genes at the expense of competing genes contained in other "survival machines."

In the real world, many organisms demonstrate altruistic behaviors; we do not mean conscious motives under this definition. The most conspicuous acts of altruism are demonstrated by parents, especially mothers, towards their children. They incubate them, feed them at enormous cost to themselves, and take great risks to protect them from predators.

Many small birds, when they see a flying predator such as a hawk, give a characteristic alarm call that causes the whole flock to take proper evasive action. The bird that gives the alarm call puts itself in exceptional danger, because it attracts the predator's attention particularly to itself.

Another example is the stinging behavior of worker bees that provides a very effective defense against honey robbers. The bees that sting the intruders are kamikaze fighters. In the act of stinging, vital internal organs are usually torn from the body and the bee dies soon afterwards. Its suicide mission may have saved the colony's vital food stocks, but the bee itself is gone.

Altruistic behaviors are widespread, and observable across species, which are determined by altruistic genes. These genes might store the code to make proteins or regulate other genes. These proteins may have one or more roles to play in forming structure, regulating functions, transporting substances, fighting infections, or acting as enzymes.

Dawkins' idea of the selfish gene was to banish any sort of altruism. According to him, no true altruism existed in nature. Apparent altruism and cooperation are the result of an individual's "selfish" genes, whose only concern is to perpetuate them. Saving the lives of close relatives can keep most of an individual's genes in circulation, even if the purpose of death is not directly related with self interest.

Even in the group of altruists, there will almost certainly be a dissenting minority who refuse to make any sacrifice. It there Is just one selfish rebel, prepared to exploit the altruism of the rest~ then he, by definition, is more likely than they are to survive and have children. Each of these children will tend to inherit his selfish traits. After several generations of this natural selection, the 'altruistic group' will be overcome by selfish individuals, and will be indistinguishable from the selfish group.

Even if we grant the improbable chance existence initially of pure altruistic groups without any rebels, it is very difficult to see what is to stop selfish individuals migrating in from neighbouring selfish groups, and, by inter-marriage, contaminating the purity of the altruistic groups. (Dawkins, 1976).

In another noted Darwinists' words, natural selection cannot lead to the establishment of altruism.

How could such altruism toward outsiders have become established in the human species? Could natural selection be invoked? This has often been tried, but not very successfully. It is difficult to construct a scenario in which benevolent behavior toward competitors and enemies could be rewarded by natural selection. (Mayr, 2001a).

However, I have an opposing opinion. It is undeniable that all individuals in any organism have some selfish genes and behaviors. However, this selfishness can coexist with altruistic acts. All organisms are a composite of selfish and altruistic behavior and genes.

In the end, natural selection will favor individuals with more altruistic characteristics because they will likely have more progenies. Two reasons dictate this. Altruistic acts might mean more risks for the actor; however, the related recipient with their shared genes survives. Even some self-sacrificing individuals would go to extinct; majority of them will survive, they would be more likely to care for their offspring and would more acceptable to their mates. Their offspring will have inherited the altruistic genes. In contract, selfish individuals would not sacrifice themselves to defend others and invest less energy in their offspring. Moreover, they would be less attractive to their mates because of their selfishness.

8-4. The survival of the most diversified

In 1798, Thomas Malthus (1776-1843) put forward his article *Essay on the Principles of Population*. He claimed that human populations would double every 25 years, however, the increase of food supply was much slower and could not meet the demands of human growth. He predicted that strife and famine would occur as the rate of population growth exceeded available resources.

Herbert Spencer (1820-1903) first introduced the term survival of the fittest. His book *Social Statics* (1850) is the pre-Darwinian prototype of the view that modern society should entail multiplication of the unfit. Darwin was profoundly affected by this idea and realized that all plants and animals had the potential to rapidly increase their numbers; however, food supplies were limited, as well as other resources fundamental for survival.

This idea that the fit individuals in a population are the ones that are least likely to die of starvation and, consequently, are most likely to pass on their traits to the next generation was crucial to his theory of natural selection. Darwin made several points in his theory:

The natural resources of an environment are restricted. Organism may have more offspring than can possibly survive. Members of a species must compete for limited resources for survival. Each members of a species has different traits, some traits enhance survival while others do not.

Organisms that have traits more useful reproduce in greater numbers. On the other hand, organisms with less favorable traits would have less offspring and eventually die out. Thus, the fittest survives. Nature selects different traits at different times. A new species would be created within a species that have the most favored characters.

Beside the erroneous idea that natural selection leads to speciation, Darwinism also conjures up an image of humans in the grip of a brutal struggle for existence. Social Darwinists advocate the elimination of the unfit as a necessary step toward progress of the society. This image is frequently linked with the cutthroat competition that arises within *laissez-faire* capitalism.

If nature achieves progress through individual competition, then the survival of the fittest must be the key to economic and social progress for humans. The struggle may be not only among individuals, but also among nations, the idea provided a justification for imperialism, colonialism, and slavery.

Most evolutionary biologists don't even want to think about the degree to which Darwinism contributed to the development of racist ideologies in the modern world. They don't really deal with the historical fact that Darwin and Galton accepted the concepts of superior and inferior races, and that Galton was particularly concerned to document the inferiority of the "Negro" and the Australian aborigine. Ernest Haeckel, one of the leading German Victorian evolutionary biologists, paved the way for the development of the elaborate system of German racism that was to develop in modern times. Evolutionary biology and racist ideology went, for a time, hand-in-hand (Rose, 1998b).

Both Darwin and Malthus were wrong in their first assumption. The natural resources of an environment are not unlimited, but abundant enough to organisms that need it. Organism die of predation, disease, or accidents, but rarely of overpopulation and the lack of food. The result of the struggle is not usually a life or death situation, in which only the winners survive and the losers die. For humans, the poorer or weaker reproduces more than the richer, or stronger.

The historical fact is that agricultural productivity has more than kept pace with the exponential growth of the human population. Consider grain production. The United States produced 378 million bushels of grain in 1839, at the start of the modern era, when its population was about 17 million. In 1957, the U.S. produced 3,422,000 million bushels, with a population of about 180 million. So while population size increased tenfold, agricultural production increased about one thousand times faster! Malthus was almost entirely in error. Agricultural production could be increased far faster than human population growth (Rose, 1998a).

Only at extreme situation is food supply the limiting factor; flood, drought, and pest can cause famine with short of food supply, however those situations occur not quite often, and only at limited geographical locations if it happens. Malthus distorted the fact by magnifying the situation and presenting it as a normal, regular case. Darwin accepted the assumption, and developed his idea of struggle for existence.

Nature has contradictory forces. It creates and eliminates variations. Nature can create variations within a species by a non-gross mutation in which numerous variants are generated. It also can create new species by the gross mutation under the GMCMI model.

Nature eliminates certain individuals within a species based on their traits, that is the traditional meaning of natural selection according to Darwin. Nature could also purge a whole species known as species selection. Nature rarely eradicates a whole species at once. It could do that easily at the initial phases of speciation, as number of members is small, the genetic makeup is similar, and the geographical location is narrow. With proliferation, the populated members will move into different geographical locations, and their genetic makeup would become varied, and they adapt to different environments. Therefore, only dramatic and widespread changes can eliminate an entire species instantly.

However, Nature could complete the elimination process gradually. Nature selection would make certain individuals more reproductive than others. The outcome of the natural selection is reduction of the diversity with a narrower genetic spectrum. If environments change, a species that appears fit then become unfit and to be eliminated soon. Thus, it is the best interest for any species to be as diverse as possible. When change occurs, some members may be affected, while others will still be well adjusted.

A species will survive longer if its members must consciously or unconsciously maintain and promote its own diversity. The larger number of individuals within a species, the more diversified species. This is one reason why many organisms demonstrate altruistic behaviors, which increases their biological diversity, even with the loss of some individuals.

Currently, at least a few hundreds species become extinct yearly. Those extinct species had undergone natural selection and they were the winner in the process of survival of the fittest according to Darwin's idea. However, such fitness did not save them from extinction. A primary cause for extinctions is change in environments. Species can no longer fit the changed surroundings. Nature might keep a few members from the extinct at some times, but the numbers were be too small to maintain a viable population.

Human society has invented rules that keep our overwhelming desires for self-preservation and self-gratification in check. The result of this practice is to maintain human diversity and to prevent one group from over-reproducing.

During an era of industrial robber barons, militarism, *laissez-faire* economics and colonial exploitation, social Darwinism became a convenient excuse for ruthlessness. Kropotkin was a Russian prince of pre-revolutionary era who railed against the social system that conferred him tremendous privilege. At a young age, he worked in Siberia and northeast China, where he observed horses forming defensive rings to resist wolf attacks, the cooperative hunting strategies of wolves themselves, and the social colonies of insects and birds. He became convinced that survival depended more on cooperation than on competition. Kropotkin published a remarkable book *Mutual Aid* (Kropotkin and Huxley, 1972), in which he was very critical of the popular view of a "struggle for existence."

Human societies have evolved from salvage to civilization. If the struggle for existence were the deciding force of nature to guide primitive human society, probably all humans would have been eliminated in the struggle among themselves. However, from a few members to the current population of 6 billion in just 200,000 years, only cooperation for existence could make such a miracle happen.

8-5. Morality and its biological origin

Christianity has deeply influenced Western morality. People believed that human were created by God in His image and have to live according to the morality He teaches. Since the publication of the *Origin*, people have begun to think if men came into being by chance and descended from the apes that developed to fight for survival, and then no reason exists to expect humans to behave in a godlike, moral fashion. We would therefore be expected to behave like "animals".

Shortly after the publication of Darwin's *Origin*, Romanes recognized what the theory was leading to and tried to warn Darwin and other against the terrible consequences, if such a theory were to become widely accepted. (Romanes, 1878). He wrote

"Never in the history of man has so terrific a calamity befallen the race as that which all who look may now behold advancing as a deluge, black with destruction, resistless in might, uprooting our most cherished hopes, engulfing our most precious creed, and burying our highest life in mindless."

Herbert Spencer introduced evolutionary modes and morality into social fields with destructive effects on the 20th century. Spencer was the one who initially invented the term evolution. He urged that the unfit be eliminated for society to evolve properly.

William Sumner (1840-1910), a professor at Yale University applied evolutionary principles to political economics. He taught many of America's businessmen and industrial leaders that the strong should succeed, the weak should perish, and helping the unfit was to injure the fit and accomplish nothing for society. According to him, millionaires were the "fittest", and modern *laissez-faire* capitalism was the necessary process to achieve that goal (Ostrander, 1971).

In its extreme form of radical thinking, Darwinism came to symbolize a new system of values derived from the evolutionary philosophy of Herbert Spencer. Universal progress was seen as a necessary outcome of the mechanical operations of the laws of nature. Humans were a product of the evolutionary process, and in the absence of any transcendental source of moral values, they would have to behave with an ethics based on the guidance of nature itself (Greene, 1981).

Edward Wilson, a professor at Harvard University and founder of sociobiology, extends Darwinian insights about bodies to behavior. He may be remembered as having revived the old controversy about nature and nurture. He defined sociobiology as "the systematic study of the biological basis of all social behavior." Much of Wilson's work focused non-human behavior and was relatively uncontroversial; His research on social insects, fish schools, birds, elephants, and carnivores was well received. However, the final chapter on human behavior "ignited the most tumultuous academic controversy of the 1970s".

When people try to derive an ethical system to judge what right and wrong, based on Darwin's theory, such a move from biology to morality always has been highly problematic. Assuming the survival of the fittest and struggle for existence were the laws, any thing that promotes own survival would be correct. A murderer would biologically correct, if he thought his action was for his/her survival and reproduction.

Shortly after publication of *the Origin of Species,* Darwin wrote to Lyell, "I have noted in a Manchester newspaper a rather good squib, showing that I have proved "might is right" and therefore Napoleon is right and every cheating tradesman is also right".

As early as 1963, the geneticist Theodosius Dobzhansky had stated the older view of human behavior. "Culture is not inherited through genes, it is acquired by learning from other human beings. In a sense, human genes have surrendered their primacy in human evolution to an entirely new, nonbiological or superorganic agent, culture."

Because of this dilemma, several evolutionists believe that morality should be disassociated from biology. Niles Eldredge, a well-known evolutionist, has written

Well, if evolution can prompt ethical systems of ruthless competition in some minds, and Christian-like harmony in others, what are we to con-dude? Here is what I have long thought: *there is no one-to-one correlation between any principle of science and any system of human behavior. In particular, there is no necessary set of ethical implications implicit in the very idea of evolution—or emanating from any subset of evolutionary theory.* To those who say there are moral lessons and ethical systems—

evil or good—implicit in the very idea of evolution, I say, A PLAGUE ON BOTH YOUR HOUSES (Eldredge, 2000).

Humanity is our nature. How was nature hard-wired into our brain under the Darwinian natural selection? How do we derive humanity from the biology? Is there any natural law to make us humane?

Darwin tried to solve this problem by appealing to an idea, which corresponded roughly to what is now called "group selection." In animals, parents must care for the young, the family group will become important and selection may well favor the instinct to defend the group and preserve the offspring. Darwin argued that in humans such instincts have been extended to include the willingness of the individual to work for the good of the tribal group. There would be a struggle for existence among the groups themselves, in which the tribes with strong cooperative instincts would eliminate the not strongly united.

Darwin himself consistently opted for selection that acted at the individual level, whereas the Darwinian explanation requires that some reproductive benefit be conferred by the behavior among groups or kin rather than individuals.

William Hamilton pioneered one of the most important concepts of sociobiology, kin selection. An individual can influence the course of selection not only by self-reproduction, but also by help of his relatives, even if he will be placed in danger. Because they share some genes, their triumph will ensure that the relative is present in the next generation (Hamilton, 1964).

The point is not whether we should associate a natural law with morality, but what is the natural law applied? If the law is the survival of the fittest and struggle for existence, then all actions are justified, as long as the act will contribute to individuals in reproduction. If kin selection is preferred, then nepotism is vindicated. Group selection would permit racism, colonialism, and slavery along the evolutionary pathway.

The implication of the GMCMI model suggests a biological rule consistent with humanity: that is the survival of the most diversified and cooperation for existence discussed in the previous section. Good acts should maintain and promote human diversity, and the converse is true. Altruistic behavior should apply to genetic relatives or our social groups, as well as to all members in human species, as they are contributive to maintaining human diversity.

The next question is how the morality entrenched in humans. The answer is by the natural selection. Morality is based altruistic behavior, which is the external demonstration of the altruistic genes imbedded inside our gene pool. Individuals with more moral behavior will have more offspring than those who do not behave in this manner.

Chapter 9

Confirmations and Falsifications

Science is more complicated than empiricism. Simple observation, repetition, and measurement are merely parts of the process. They could be important initial steps toward scientific discovery; only they are interpreted, tested, and used properly. The scientific method is a program for research comprised of four main steps. These four steps are:

1. Observe the facts
2. Form a unifying hypothesis to explain these facts
3. Deduce predictions from the hypothesis
4. Search for confirmations of the predictions. If they are false then return to step 2

The real test of any scientific theory is its ability to generate testable predictions and have them proven. Any proposed model can be used to deduce predictions. These predictions are then compared with the real world to determine how consistent the model is with verifiable evidences. If, without assuming the model is true, many facts do not support these predictions; thus, the model should be considered incorrect.

If a theory resists the test of time and the compilation of new data, it will be considered a scientific fact. A theory will be regarded as scientific "fact", if it is well-supported by many independent evidences (Kuhn, 1962). As Stephen Gould stated, a scientific fact is not "absolute certainty," but simply a theory that has been "confirmed to such a degree that it would be perverse to withhold provisional consent."

Scientists are constantly making new observations. New observations, although not predicted, should be explained retrospectively by the theory; otherwise, they can count as falsifications. New information, especially details of a process previously not understood, can impose new limits on the original hypothesis.

Testability lies at the heart of the scientific enterprise. If we make a statement about that state of things, for instance, how the Earth originated and how the world's species came into being—we must be able to make *predictions* about our observations of the world if that statement is true. If theories survive repeated testing, we should tentatively consider them as true statements. Repeated failure to confirm predicted observations means an idea should be abandoned, no matter how it was respected.

The issue of finding a criterion for distinguishing science from pseudoscience is always controversial. Karl Popper, an Austrian-born British philosopher of natural and social science, is a pivotal person in understanding the differences between science and pseudoscience. Popper believes that the criterion of the scientific status of a theory is its falsifiability, i.e. person who establishes a theory should specify what happen will make theory wrong. If a theory can justify any outcome, it is not a scientific one.

The Marxist's theory of capitalism was very popular at the beginning of the 20th century, which could explain very outcome of capitalist society. Marx predicted that wages would fall as capitalists exploited workers, while an increase in wages was also consistent with Marxism; as capitalists propped up the system with bribery. Therefore, Marx's theory of capitalism could not be considered scientific by Popperian criteria, as it can justify any outcome of the capitalist society.

One of the key characteristics of science is the concept of the testable hypothesis; therefore, independent observers must ratify the predictions. By testable, we mean the predictions must include examples of what *should* be observed and what *should not* be observed, if the hypothesis were true.

In principle, a scientific hypothesis should rule out conceivable possibilities, which is the essence of the Popperian falsifiability criterion (Popper, 1968). For instance, a hypothesis that real life is a reflection of past life is not a scientific one, as it is impossible to determine past life by scientific methods. In contrast, Einstein's theory of mass energy is a scientific one, as one can establish the relationship between mass and energy using reliable scientific methods. If observations were incompatible with the predictions of Einstein's theory, the theory is falsifiable.

According to Karl Popper, scientific theory with explanatory powers should make risky predications. Successful predications are impressive only to the extent that failure was real prospect.

In this chapter, I would predict four outcomes. Others remain for future study. Following conventions are applied to these predictions:

1) *Prediction:* prediction is used as to what should be observed, if the model is correct.

2) *Confirmation:* confirmation is used if actual discoveries are presented for each prediction.

3) *Potential confirmation:* potential confirmation will be used, if actual data are not available, but can be obtained through scientific efforts.

4) *Potential falsification*: potential falsification is used for observation to be seen, if the theory were not correct.

With development of biological research, these predictions will be further substantiated and confirmed.

9-1: Every species has two 'Eves'
9-2: Homogeneity of initial genetic structure
9-3: Terminal species
9-4: Inconsistencies of molecular phylogeny

9-1. Every species has two "Eves"
Prediction

According to the GMCMI model, every species have two Eves. The first one is the single ancestor mother, who gave birth to a new species, the second "Eve" or 'Eves" is a group of females with very similar genetically structure, who are the first generation or "seed" of new species. In terms of human, most likely, its ancestor or "first Eve" was one member of ape-like animals; the second "Eves" was a group of mothers with identical human genetic structure and phenotype.

Confirmation

In 1987, University of California geneticist Allan Wilson announced the results of a study utilizing mitochondrial DNA as a marker to trace the ancestry of modern humans. They believed the family tree led to a woman who lived in Africa about 200,000 years ago (Cann, Stoneking et al., 1987). Found outside the cell nucleus and only traceable through the female line only, mtDNA is "extra" genetic material. These researchers collected and analyzed mtDNA material from 147 women from Africa, Asia, Europe, Australia, and New Guinea. Despite similarity, data fit two main groupings: Africa and everywhere else. Wilson and colleagues believed that their findings supports the conclusion that "the common ancestor of modem humans lived in Africa, about 200,000 years ago," which was relatively new by the evolutionary timescale.

However, some traditional anthropologists argued that fossil evidence does not support a recent African origin for all the world's people. In 1992, geneticist Alan Templeton of Washington University to take a fresh look at how the Berkeley team reached their conclusions. First, Templeton found that the original "Eve" researchers had used their computer programs improperly. Their mtDNA analysis had generated a parsimonious family tree, but there could be millions of equally parsimonious family trees depending on the programming algorithm. A tree with African roots is not any more probable than one with Asian or European roots. "The inference that the tree of humankind is rooted in Africa is not supported by the [genetic] data," Templeton wrote in the journal *Science*.

The proposed model predicts the correctness of "Eve theory"; whether or not the original location is Africa. Not only humans, but also all species, have their own Eves. Duplicating mtDNA studies in other animals can examine this idea.

Potential Falsification:

If we could confirm multiple origins of geographical locations for any species, it would be catastrophically problematic for the GMCMI model.

9-2: Similarity of the initial genetic structure
Prediction:

By natural selection, at the beginning of any new species, diversity of genetic structures within the species existed, whereas in the GMCMI model, all species had similarity genetic structure at the beginning.

Potential Confirmation:

The cheetah's range once spanned the globe. Nearly all cheetahs today are almost genetically identical. Also, consider the Humboldt gopher and skin grafts, which was detailed in earlier. The author speculated that they might be relatively new evolved species, and the DNA fragments responsible for immunity lack flexibility, so they are still similar to their original ones.

Potential Falsification:

If we could confirm diversity in DNA structure in the initial members of any species, the GMCMI model would be shattered.

9-3: Terminal Species
Prediction:

As discussed in the chapter 5, the ability to generate prolific novelties in any species is directly related to the number of fraternal births. The size of animals throughout evolution has increased, while the number of siblings in the same birth decreased. Many species only have a singleton birth, such as humans, elephants, and tigers; multiple siblings from the same births are only the exception.

The author proposes a concept, called 'terminal species', which is defined as 'a species that loses natural ability to generate a novel species because of low number of siblings from the birth'. A low number of births make speciation very difficult or impossible in such an animal. All singleton birth animals, including human beings, are terminal species.

With the trend that multiple births decrease and body size increases, it is expected that the speed of speciation will decrease to the point where it stops.

Confirmation:

Several authors suggested the evolution might be a self-limiting process:

"Evolution is thus seen as a series of blind alleys. Some are extremely short - those leading to new genera and species that either remain stable or become extinct. Others are longer - the lines of adaptive isolation within a group such as a class or subclass, which run for tens of millions of years before coming up against their terminal blank wall. Others are still longer - the links that in the past led to the development of the major phyla and their highest representatives; their course is to be reckoned not in tens but in hundreds of millions of years. Even the ants and bees have made no advance since the Oligocene. For the birds, the Miocene marked the end; for the mammals, the Pliocene." (Huxley, 1963).

"In Eocene times - say between 50,000,000 and 30,000,000 years ago-small primitive mammals rather suddenly gave rise to over a dozen very different Orders - hoofed mammals, odd-toed and even-toed, elephants, carnivores, whales, rodents, bats and monkeys. And after this there were no more Orders of mammals ever evolved. There were great varieties of evolution in the Orders that had appeared, but strangely enough Nature seemed incapable of forming any new Orders...No new types of fishes, no groups of molluscs, or worms or starfishes, no new groups even of insects appear to have been evolved in these latter 30,000,000 years." (Broom, 1951).

Currently, no new species evolving from a singleton animal has been confirmed.

Potential Falsification:
We should never find a species with a singleton birth as its ancestor. Any finding of a new species evolving from terminal species would be inconsistent with the GMCMI model.

9-4: Inconsistency of molecular phylogeny
Prediction:
A phylogeny is the evolutionary history of a group of organisms. Before the 1960s, researchers could only deduce the relatedness of organisms by comparing their anatomy or physiology. Genomic science and molecular technology made it possible to document molecular evolutionary history by studying genomic sequences from prokaryotes, yeast, and humans

According the GMCMI model, speciation is the outcome of gross mutations by a random process. There are no specific patterns how, why, where the mutations occur, any genes could be involved in the gross mutation. Irregularity of involvement of genes would make the inconsistency of results in molecular phylogeny a rule, not an exception.

Confirmation
In the mid-1960s, Emile Zuckerkandl and Linus Pauling of the California Institute of Technology initiated a strategy to study evolution in selected genes of proteins (Zuckerkandl and Pauling, 1965). Most of the early work in molecular phylogeny relied on

proteins. With more information available from sequencing, it became more common to analyze DNA and rRNA sequences, instead of the proteins. Since 1980, the DNA sequences that code for rRNA have provided important data for molecular phylogeny.

It was expected that phylogenetic trees should be approximately the same regardless of which molecules are chosen for comparison, but that hypothesis fell apart when more data became available.

A variety of genes from different organisms demonstrated that their relationships contradicted the evolutionary tree of life derived from rRNA analysis. Different molecules lead to different phylogenetic trees. According to biologist Carl Woese, an early pioneer in constructing rRNA-based phylogenetic trees, "No consistent organismal phylogeny has emerged from the many individual protein phylogenies so far produced. Phylogenetic incongruities can be seen everywhere in the universal tree, from its root to the major branchings within and among the various [groups] to the makeup of the primary groupings themselves."

Potential Falsification:

If we would find consistency among genes by molecular phylogeny techniques, the GMCMI model would suffer greatly.

Chapter 10

Summary

This chapter will summarize the proposed mechanisms and highlight major points. Readers may return to the relevant chapters for more detail information.

10-1. Timing of mutations
10-2. Inbreeding among MISTWGM
10-3. Factors to aid proliferation and selection
10-4. Characteristics of gross mutations
10-5. Types of speciation
10-6. Challenges and their answers
10-7. Moral implications of the model
10-8. The GMCMI model has many independent
 evidences
10-9. Science or belief?

10-1. Timing of mutations
Mutations are sources of variation and the timing of mutation is very important in speciation. Although numerous pathways for the gross mutation exist, the zygote is the most likely timing for the mutation occurs to cause speciation.

10-2. Inbreeding among MISTWGM
Speciation by chromosomal rearrangements is not a new idea. The most common challenge for the

idea is how the newcomers overcome reproductive barriers and develop into novelties with reproductive capability.

The GMCMI model includes a rational biological process to solve the critical issue: new species does not come into the world as a singleton, but as a cluster or group, a group in opposite sexes with the same properties. The process is as follow:

At the beginning, a mutation occurred on two zygotes in the different genders. The mutant zygotes are the first cells of new species that self-duplicate to become MMIZWGM, who developed and were born as MISTWGM. Depending on animal, the number of siblings in MISTWGM could be a few to hundreds. They were born in the same birth, live together, and have the same genetic structure, inbreeding among MISTWGM naturally occurs. Few, if any, reproductive barriers prevent brothers and sisters having reproductive offspring.

10-3. Factors to aid reproduction and selection

The process in animals, which uses viviparous animals to illustrate the process, is more complicated than one in plants. Animals migrate; factors that affect their successful reproduction are multi-faceted. Several factors have been suggested for their roles in either reproduction or natural selection of new species.

The critical factor to limit a successful speciation is the reproduction of PSN. If PSN were born in MISTWGM, reproduction of the next generations would certainly become possible. Any species with singleton or a low number of offspring siblings would make it impossible to have further speciation, they are a terminal species. Humans, tigers, and elephants are all terminal species; no other species could be evolved from them.

Viable MISTWGM can inbreed among themselves, only if they have external, recognizable, and stable features. Novelties without SMRS cannot recognize each other and pair successfully.

New species was generated by chance under the GMCMI model; particular features would make them to have better opportunity to survive when competing with other species.

Increased body size in new species is often associated with strength and physical endurance, which provides a new species with more advantage for survival under the natural selection.

If a new species has new features than their ancestral species, they might move into a new theater for living with less competition.

The initial 'seed' of any species is a pure line; which means that they all have identical genotypes and phenotypes. Genetic structures in those seeds will make some species diversify more easily than other ones. Animals with poor genetic diversity do not have good resistance to infection. If a new disease arises, they could be eliminated entirely.

10-4. Characteristics of mutations

According to the GMCMI model, a mutation can be either non-gross (micromutation) or gross one (macromutation). The non-gross mutation will generate diversity within species, whereas the gross one might lead to generation of PSN or MISTWGM. Both types of mutations are random processes. Accumulation of micromutation would never lead to macromutation.

10-5. Types of speciation

In many textbooks, three types of geographical speciation are often discussed and assumed to be true:

instantaneous, sympatric and allopatric, so far, only instantaneous speciation, or polyploid is the confirmed case.

By the GMCMI model, all speciation is instantaneous speciation; it is also sympatric (the same location with ancestor) and cladogenetic. The allopatric and anagenesis speciation are just illusions.

10-6. Challenges and their answers

The GMCMI model is a new hypothesis of the speciation process with sound biological explanations, with which almost all of inexplicable under the Darwin's theory have biologically plausible answers. Roughly, they are grouped into five types:

1. Naturalists have some sort of answers, such as the lack of transitional links in fossils, albeit unconvincingly. The GMCMI model provides more plausible ones.
2. Naturalists do not have plausible answers, while the GMCMI model does.
3. Questions can be only answered by GMCMI, even partially.
4. Similar answers by both mechanisms.
5. No discussion by naturalists.

These questions have been discussed in chapters 5, 6, and as well as 7, they are categorized in Table 10.1.

Table 10.1. Classifications of Evolution Issues

Class	Evolution Issues
1	Transitional links, vestigial organs, atavisms, magical organs, progression and trend, industrial melanism, organs and environments, cladogenesis and anagenesis, geographical isolation, drug resistant bacteria, bottle-neck effect
2	Genetic mechanism for alteration of chromosomal structures, chicken-egg paradox, overcome of reproductive barriers, mosaic of evolution, lateral transfer, rates of evolutionary change
3	Cambrian exploration
4	Geographical radiation, variation and natural selection within speciation
5	Human Infertility with abnormal karyotype

10-7. Moral implications of the model

Selections work on multiple levels: by traits, groups, and chance. According to Darwinians, only selection by traits is considered as natural selection, because it can alter the frequencies of phenotypes.

Survival of the fittest does not mean the survivor must have better traits than ones who did not. An individual can survive just due to chance, or in a particular group. The survival of the fittest might be correct statistically, and it only applies to a population, not individuals, and it is also just relative, and temporary.

Additional to selfish genes proposed by Richard Dawkins, organisms also contain genes for altruistic behaviors. Natural selection would favor individuals with more altruistic characters. Altruistic genes are also the foundation for morality. Individuals with altruistic characters would be more likely to spend more time to take care of their offspring, and more acceptable by their partners, and their children will tend to have the genes, whereas selfish individuals do not sacrifice themselves in defending other peers and their kins, and they would probably have less offspring.

Darwinians mistakenly held that natural resources are restricted and members of a species must compete for limited resources to survive, and individuals with better traits will have more progenies, this might be true in a very limited sense. The natural resources of an environment are relatively abundant to organisms it feeds at most of the time, most animals died of predators, disease, accident, not of lack of food.

The inference of the model implies a new biological rule for humanity. Good acts should be valuable to maintain and promote human diversity.

10-8. The GMCMI model has many independent evidences

Species differs in their genetics, morphologies, and behaviors. The chromosomes of many taxa have been described and species are known to differ in their chromosomal numbers, lengths, and structure. Humans, for example, have 46 chromosomes, whereas chimps, gorillas, and the orangutan have 48 chromosomes, suggesting that a pair of chromosomes have fused in our ancestry.

Many distinguished predecessors have reached the idea of speciation by macromutation independently. For instance, Broom and Schindewolf reached the conclusion that the individual is the unit of evolutionary change. Berg (Berg, 1969) reached the same conclusion as Punnett (Punnett, 1915), Osborn, and Bateson (Bateson, 1913) realized that the role of selection is to eliminate variations rather than to produce them. Grassé came to similar finding (Grassé, 1977). Goldschmidt, Schindewolf, and Berg all concluded that saltation is the mechanism for speciation.

In the chapter 9, I discussed the mitochondrial "Eve" and results from the studies of molecular phylogeny, the evidences from these studies strongly support the model from the point of views of molecular biology.

A few months before this book was published, the chapter 5 of the book was sent to several well-known experts for their opinions. Dr. Jonathan Wells, a well-known ID advocate and the author of *Icons of Evolution,* replied,

"The real test of your theory, of course, will be the evidence. As I'm sure you know, other biologists have proposed chromosomal-rearrangement theories of speciation, but it's my understanding that except for polyploidy in plants such proposals have lacked sufficient persuasive evidence to convince a majority of evolutionary biologists. On this point, I hope you can succeed where others have failed (personal communication)"

As I discussed in the previous chapter, the model solves the major challenge facing macromutation theory: that is how a new species generated by gross mutation have their reproductive offspring. The biological processes detailed in the chapter have never been suggested in any previous proposals.

In a separate communication, Dr. Michael Behe, the author of *Darwin's Black Box*, wrote:

"Your proposal sounds interesting, but strikes me as rather vague. I think you should add more documentation, if there is any, that MIST is possible for animals, especially vertebrates. Your examples of the cheetah and pocket gopher are interesting, but speculative (personal communication)".

I conjecture that he referred to the evidence in animals. In the chapter 5, I discuss cases of instantaneous speciation in plants and two genetic identical animals as the potential candidates for the direct evidences. Admittedly, compared with plants, the evidence in animals is scarce. However, its paucity, like the lack of intermediates in fossils records, might be considered a defense of the hypothesis. Animals are constantly migrating to new locations for food and avoidance of predators, which would make the evidence very rare. One cannot imagine a whole group of identical animals voluntarily staying together for generations and waiting. They must simply be retarded and do not survive. The scenarios of the cheetah and gopher are speculative. I present them as alternative explanations to the bottleneck effect to explain why they are found to be identical populations. Even it is confirmed later that my explanations are incorrect, the proposed model still holds, as long as other evidences can be established with scientific studies.

Scientific evidence does not arise by itself. It is drawn from hypothesis-driven research. Like industry, designing always precedes products. Since no one has used the idea in the model as the basis of designing research, it is not surprising that only a few such products are found. They are the by-products of other research.

The most convincing evidence in favor of the GMCMI model is the explanatory power associated with it. Numerous mysteries in biology would have plausible answers, only if one assumes the validity of the hypothesis. Given the chicken and egg paradox, mosaic of evolution, lateral transfer, rates of evolutionary change, and many others, it would be inconceivable for any other evolutionary theories to answers many these puzzles with such a simple and plausible idea.

10-9. Science or belief?

The speciation by natural selection is considered a "second-order theory" in the J. L. Monod essay:

'The other great difficulty about the theory of evolution is that it is what one might call a second-order theory. Second-order, because it is a theory aimed at accounting for a phenomenon that has never been observed, and that never will be observed, namely evolution itself. In the laboratory we are able to set up conditions so that we may be able to isolate mutations of a given bacterial strain, for instance; but to observe a mutation is a very far cry from observing actual evolution. That has never been observed even in its simplest form—the one that is required by the modem theorists to account for evolution, namely the simple differentiation of one species from another. This is a phenomenon that has never been seen. I would not say it never will be, but it seems extremely doubtful (Monod, 1997)".

The theory of natural selection has to be taken for granted based on imagination and belief, which does not differ from religious beliefs. According to certain religious, there is eternal life after death; no one has any scientific evidence to show what happens after death. To believe apes became humans under natural selection after a million years, one must be braver and willing to turn a blind eye, as evidence is stacked against it.

To solve scientific problems, one should use known mechanisms, parameters, and factors to explain unknown phenomena. For example, we do not know the etiology of severe acute respiratory syndrome (SARS), we would investigate all possible pathogens until *Coronavirus* is determined to be the infectious agent. It is far from enough by speculating that agent X exists, which would cause all the symptoms, if we do not know what agent X is

In the natural selection, there are the four elements in speciation process: mutations, survival of the fittest, environmental change, and a lengthy process. The first three elements are understandable, and testable; however, the time scale makes the fourth element untestable and untenable for scientific investigation, it is a belief and speculation. Thus far, no a single exclusive evidence has been found to support the theory of natural theory despite many years of efforts. The entire presumptive evidences can be explained by the alternative mechanism, and only even better.

The idea of Punctuated Equilibria has a similar drawback. It provides paleontologists with an explanation for the patterns in the fossil record by assuming that subgroup of a species at a peripheral location might undergo dramatic changes for unknown reasons with an unclear mechanism. In essence, it attempts to explain the speciation process with another set of speculations, and does not solve any challenge raised in Chapter 4 and appendix I, the all of major issues with Darwin's theory remain completely unexplained, with exception of the missing fossil links.

The GMCMI model has five elements: the zygotes, the gross mutation, self-replication, supertwins, and inbreeding. These observable phenomena occur at any time and place. The challenge is not whether speciation happens by the proposed model, but how to detect these new species at the initial stage.

The concepts of the GMCMI model are easily understood and intuitive. No of these elements is based on belief or speculation. They do not require advanced scientific knowledge, and they are often included in textbooks for high school students.

William of Occam was a renowned scholastic philosopher of the fourteenth century. He is best known for the scientific tenet "Occam's Razor". It states, "Entities are not to be multiplied without necessity." In other words, a well-constructed theory about nature is the simplest possible explanation consistent with the facts. The "razor" shaves off any unnecessary superfluities or complications.

"Applied to geology and paleontology over the years, "Occam's razor" has raised (and resolved) key questions: Why postulate that God continually intervened in changes of climate, variations in species, floods, etc., when it is simpler to assume establishment of uniform laws (proximate causes) that operate to produce all subsequent events? Why assume God made fossils of creatures that never existed in order to test man's faith, when there is the much simpler explanation that they are the remains of animals that once lived?

And why assume (as Charles Darwin did) that patterns of apparently long stability and abrupt change are functions of "imperfections" in the fossil record, which may someday be corrected by new discoveries? The simpler explanation is that these widespread geological patterns reflect actual events (punctuated equilibrium) (Milner, 1990)."

When applying the same logic to the GMCMI model, why assume the mechanism natural selection is correct and search for missing links? The simpler explanation is to assume the GMCMI model correct, which predicts that no such links exist, period.

In addition, why postulate that EP can provide a proper explanation for speciation, even missing links get a sound explanation. EP cannot explain the rate change, chicken-egg paradox, and reproductive barriers, just name a few, when the GMCMI model can satisfy all those remaining issues.

This book is not intended to complete the story, but it is to try to initiate a new avenue of thinking. The model should be the paradigm to stimulate a new wave of scientific research. Research on Darwin's theory did not generate any confirmation in the over past 140 years and its continuation seems a misuse of scientific resources.

Considering there might have been over 100 million species formed over the past 4.5 billions year, the details might need some modifications to accommodate the unlimited varieties. However, the principle of instant speciation by macromutation and mutants inbreeding should be the core in any modalities.

"Nothing in biology makes sense except in the light of evolution (Dobzhansky, 1973)". However, so many things do not make any sense, if natural selection is assumed as the correct mechanism of speciation. Almost all of the questions created by natural selection would be resolved under the GMCMI model.

Closing Marks

Evolution is the most profound and insightful idea conceived in history. It was proposed in the 17th century in Europe and developed later by Charles Darwin in *On the Origin of Species* in 1859. , He not only proposed that evolution occurred with supporting evidence, but also postulated the mechanism of natural selection as the cause of evolution. Since its publication, heated debates have raged among biologists and religionists concerning its correctness. Many agree that Darwin's theory does not provide the whole picture.

How do scientists discipline themselves to understand the natural world? With different ways to answer this question, any scientific theory must be proven with natural observations or laboratory analysis. Thus far, scientists have perplexed themselves with many natural phenomena. The conflicting points arise when one assume the natural selection is the correct answer for the mechanism of speciation.

In this book, the author proposed a novel mechanism of speciation, which marks the end to lengthy debates. Many challenges raised by the opponents of the natural selection have biologically plausible answers. Numerous examples previously unexplainable by natural selection are demystified under the GMCMI model. None of cases explained by natural selection are unexplained with the model.

Science is not a belief system as no idea is sacred. No statement is considered the ultimate truth without vigorous challenges. Niles Eldredge stated, "Scientists are supposed to hold their theories lightly, to explore all opportunities for knocking them as under, to "falsify' them." Science progresses by showing that something is wrong. If an idea squares with the facts, then it might be right. If not, then that idea is defeated. We are far more certain when an idea is wrong than when it is right".

The history of science is littered with discarded ideas—notions were not borne with extensive investigation. Philip Gold, a famous ID advocate, wrote:

"At the moment, the various paradigms provided by the scientists and allegedly scientific thinkers of the 19th and early 20th century West are failing: this is the necessary prelude to the next set of shifts. Karl Marx has been consigned to the trash compactor of history. Sigmund Freud has been composted. Albert Einstein is in trouble. (The speed of light isn't constant, and may have been exceeded recently in, of all places, New Jersey.) Of the great thinkers who fashioned the modern world-view, only one — Charles Darwin remains inviolable (Gold, 2000)"

This time, Charles Darwin is violated, however it is still the Nature that creates all species.

Appendix I

Mysteries of Mutations

In this chapter, I have abstracted several paragraphs from pages 393-459 of *Origin of the Life*, Volume Two of Evolution Disproved Series (http://www.evolution-fact.org). They are challenges to Darwin's theory from the viewpoint points of DNA mutation. Similar challenges will be raised against the proposed model. Therefore, I will briefly answer the potential challenges in advance here; many of these questions have been discussed in detail in the previous chapters. My comments are below the question in bold italics.

Why mutations cannot produce cross-species change
A mutation is damage to a single DNA unit (a gene). If it occurs in a somatic (body) gene, it only injures the individual; but if to a gametic (reproductive) gene, it will be passed on to his descendants.

A gross mutation can affect one or more chromosomes, not just a single gene. It is mutations in a zygote, not a somatic cell or a gamete that leads speciation. See chapters 1 and 5 for the detail.

"It must not be forgotten that mutation is the ultimate source of all genetic variation found in natural populations and the

only new material available for natural selection to work upon."—
E. Mayr, Populations, Species and Evolution (1970), p. 103.

Agree

"The process of mutation is the only known source of the new materials of genetic variability, and hence of evolution."—*T. Dobzansky in American Scientist, 45 (1957), p. 385.*

Agree

"The complete proof of the utilization of mutations in evolution under natural conditions has not yet been given."—
Julian Huxley, Evolution, the Modern Synthesis, pp. 183 and 205.

Agree

Overview of the situation—Mutations generally produce one of three types of changes within genes or chromosomes: (1) an alteration of DNA letter sequence in the genes, (2) gross changes in chromosomes (inversion, translocation), or (3) a change in the number of chromosomes (polyploidy, haploidy). But whatever the cause, the result is a change in genetic information.

Here are some basic hurdles that scientists must overcome in order to make mutations a success story for evolution: (1) Mutations must occur quite frequently. (2) Mutations must be beneficial—at least sometimes. (3) They must effect a dramatic enough change (involving, actually, millions of specific, purposive changes) so that one species will be transformed into another. Small changes will only damage or destroy the organism.

See comments below.

NEO-DARWINISM—(*What the Public is not Told*) When Charles Darwin wrote *Origin of the Species*, he based evolutionary transitions on natural selection. In his book, he gave many examples of this, but all his examples were merely changes within the species.

Probably true.

Finding that so-called "natural selection" accomplished no evolutionary changes, modern evolutionists moved away from Darwinism into neo-Darwinism. This is the revised teaching that it is mutations plus natural selection (not natural selection alone), which have produced all life-forms on Planet Earth.

Gross mutations, non-gross mutations, plus natural selection have produced all varieties of life on Earth.

Neo-Darwinists speculate that mutations accomplished all cross-species changes, and then natural selection afterward refined them. This, of course, assumes that mutations and natural selection are positive and purposive.

Mutations are random; but natural selection causes the mutation to appear positive and purposeful by keeping the fittest species and variants.

1 - Four Special Problems
In reality, mutations have four special qualities that are ruinous to the hopes of evolutionists:

(1). Rare Effects—Mutations are very rare. This point is not a guess but a scientific fact, observed by experts in the field. Their rarity dooms the possibility of mutational evolution to oblivion.

"It is probably fair to estimate the frequency of a majority of mutations in higher organisms between one in ten thousand and one in a million per gene per generation."—*F.J. Ayala, "Teleological Explanations in Evolutionary Biology," in Philosophy of Science, March 1970, p. 3.*

Mutations are simply too rare to have produced all the necessary traits of even one life-form, much less all the creatures that swarm on the earth.

The Studies on human infertility discussed in Chapter 7 show that about 2% of couples have abnormal karyotypes, which is the rate of the gross mutation at microscopic level The submicroscopic level is even higher, they are all gross mutations.

(2) Random Effects—Mutations are always random, and never purposive or directed. This has repeatedly been observed in actual experimentation with mutations.

"It remains true to say that we know of no way other than random mutation by which new hereditary variation comes into being, nor any process other than natural selection by which the hereditary constitution of a population changes from one generation to the next."—*C.H. Waddington, The Nature of Life (1962), p. 98.*

The mutations are completely random processes, and natural selection might create trends.

"It is our contention that if 'random' is given a serious and crucial interpretation from a probabilistic point of view, the randomness postulate is highly implausible and that an adequate scientific theory of evolution must await the discovery and elucidation of new natural laws."—*Murray Eden, "Inadequacies of Neo-Darwinian Evolution as Scientific Theory," in Mathematical Challenges to the Neo-Darwinian Theory of Evolution (1967), p. 109.*

The new natural laws are implicated in the GMCMI model.

Mutations are random, wild events that are totally uncontrollable. When a mutation occurs, it is a chance occurrence: totally unexpected and haphazard. The only thing we can predict is that it will not go outside the species and produce a new type of organism.

This we can know as a result of lengthy experiments that have involved literally hundreds of thousands of mutations on fruit flies and other small creatures.

Evolution requires purposive changes. Mutations are only chance occurrences and cannot accomplish what is needed for organic evolution.

Purposeful changes occur by natural selection after formation of a new species through the random mutations.

(3) Not Helpful—Evolution requires improvement. Mutations do not help or improve; they only weaken and injure.

"But mutations are found to be of a random nature, so far as their utility is concerned. Accordingly, the great majority of mutations, certainly well over 99%, are harmful in some way, as is to be expected of the effects of accidental occurrences." *H.J. Muller, "Radiation Damage to the Genetic Material," in American Scientist, January 1950, p. 35.*

Even 1% or less of mutations have a beneficial effect, that is sufficient to generate all species observed.

(4) Harmful Effects—(*mutations are always harmful*) Nearly all mutations are harmful. In most instances, mutations weaken or damage the organism in some way so that it (or its offspring if it is able to have any) will not long survive.

"A proportion of favorable mutations of one in a thousand does not sound much, but is probably generous, since so many mutations are lethal, preventing the organism from living at all, and the great majority of the rest throw the machinery slightly out of gear."—*Julian Huxley, Evolution in Action, p. 41.*

Mutations are rare, random, almost never an improvement, always weakening or harmful, and often fatal to the organism or its offspring.

Answer the same as above.

2 - Twenty-Eight Reasons
Here are 28 reasons why it is not possible for mutations to produce species evolution:
1 - not once—Hundreds of thousands of mutation experiments have been done, in a determined effort to prove the possibility of evolution by mutation. And this is what they learned: not once has there ever been a recorded instance of a truly beneficial mutation (one which is a known mutation, and not merely a reshuffling of latent characteristics in the genes), nor such a mutation that was permanent, passing on from one generation to another!
Read the above paragraph over a couple times. If, after millions of fruit-fly mutation experiments,

scientists have never found one helpful and non-weakening mutation that had permanent effects in offspring—then how could mutations result in worthwhile evolution?

"Mutations are more than just sudden changes in heredity; they also affect viability [ability to keep living], and, to the best of our knowledge invariably affect it adversely [they tend to result in harm or death]. Does not this fact show that mutations are really assaults on the organism's central being, its basic capacity to be a living thing?"—*C.P. Martin, "A Non-Geneticist Looks at Evolution," in American Scientist, p. 102.*

The experimental conditions may differ from the natural conditions. All of experiments were designed without knowledge of the GMCMI mode; thus, it is not surprising why scientists have never found one helpful and non-weakening mutation that had permanent effects in offspring.

2 - Only harm—The problem here is that those organisms which mutations do not outright kill are generally so weakened that they or their offspring tend to die out. Mutations, then, work the opposite of evolution. Given enough mutations, life on earth would not be strengthened and helped; it would be extinguished.

This gradual buildup of harmful mutations in the genes is called genetic load.

"The large majority of mutations, however, are harmful or even lethal to the individual in whom they are expressed. Such mutations can be regarded as introducing a 'load,' or genetic burden, into the [DNA] pool. The term 'genetic load' was first used by the late H.J. Muller, who recognized that the rate of mutations is increased by numerous agents man has introduced into his environment, notably ionizing radiation and mutagenic chemicals."—*Christopher Wills, "Genetic Load," in Scientific American, March 1970, p. 98.*

Even 1% or less of mutations are helpful enough to generate all species observed.

3 - usually eliminate—Because of their intrinsic nature, mutations greatly weaken the organism so much; so that if that organism survives, its descendants will tend to die out.

The result is a weeding-out process. Contrary to the hopes of the neo-Darwinians, natural selection does not enhance the effects of the mutation. Natural selection eliminates mutations by killing off the organism bearing them!

Same as above.

4 Mutagens—It is a well-known fact that scientists have for decades been urging the removal of radiation hazards and mutagenic chemicals (scientists call them mutagens) because of the increasing damage mutations are doing to people, animals, and plants.

It is time that the evolutionists, who praise the value of mutations, admit very real facts. How can such terrible curses, which is what mutations are, improve and beautify the race—and produce by random action all the complex structures and actions of life?

The mutations could occur under 'natural conditions' without induction by mutagenic chemicals. Speciation could occur with gross mutations such as DNA rearrangements.

5 dangerous accidents—How often do accidents help you? What is the likelihood that the next car accident you have will make you feel better than you did before?

Because of their random nature and negative effects, mutations would destroy all life on earth, were it not for the fact that in nature they rarely occur. Actually, a significant part of the grave danger in mutations is their very randomness! A mutation is a chance accident to the genes or chromosomes.

Just like the question 2 above.

6.Intertwined catastrophe—A new reason why mutations are so insidious has only recently been discovered. Geneticists discovered the answer in the genes. Instead of a certain characteristic being controlled by a certain gene, it is now known that each gene affects many characteristics, and each characteristic is affected by many genes! We have here a complicated interweaving of genetic-characteristic relationships never before imagined possible!

One step by a gross mutation may cause catastrophic reactions, which leads to hundreds of alterations in genotypes and phenotypes, predicted by the GMCMI model.

7 - only random effects—So far in this chapter, we have tended to ignore the factor of random results. What if mutations were plentiful and always with positive results, but still random as they now are? They would still be useless.

Even assuming mutations could produce those complex structures called feathers, birds would have wings on their stomachs, where they could not use them, or the wings would be upside down, without lightweight feathers, and under- or oversized.

Most animals would have no eyes, some would have one, and those that had any eyes would have them under their armpits or on the soles of their feet.

The random effects of mutations would annihilate any value they might otherwise provide.

One step by a gross mutation has hundreds of alterations in many organs; ones with coordinated changes survive, and others without would die. One would only see what has survived.

8 all affected—Mutations tend to have a widespread effect on the genes.

"Moreover, despite the fact that a mutation is a discrete, discontinuous effect of the cellular, chromosome or gene level, its effects are modified by interactions in the whole genetic system of an individual. Every character of an organism is affected by all genes, and every gene affects all other characters. It is this interaction that accounts for the closely knit functional integration of the genotype as a whole."—*Ernst Mayr, Populations, Species, and Evolution, p. 164 [emphasis his].*

Each mutation takes its toll on large numbers—even all the genes, directly or indirectly; and since 99 percent of the mutations are harmful and appear in totally random areas, they could not possibly bring about the incredible life-forms we find all about us.

Since each altered characteristic requires the combined effort of many genes, it is obvious that many genes would have to be mutated in a GOOD way to accomplish anything worthwhile. But almost no mutations are ever helpful.

Thousands and thousands of generations of fruit flies have been irradiated in the hope of producing worthwhile mutations. But only damage and death has resulted.

Just as question 7 above.

9 like throwing rocks—Trying to accomplish evolution with random, accidental, harmful mutations is like trying to improve a television set by throwing rocks at it (although I will admit that may be one of the best ways to improve the benefit you receive from your television set).

H.J. Muller won a Nobel Prize for his work in genetics and mutations. In his time, he was considered a world leader in genetics research. Here is how he describes the problem:

"It is entirely in line with the accidental nature of mutations that extensive tests have agreed in showing the vast majority of them detrimental to the organism in its job of surviving and reproducing, just as changes accidentally introduced into any artificial mechanism are predominantly harmful to its useful operation. Good ones are so rare that we can consider them all bad."—*H.J. Muller, "How Radiation Changes the Genetic Constitution," in Bulletin of Atomic Scientists, 11(1955), p. 331.*

Even submicroscopic portions of mutations were good enough to cause all species observed.

10 mathematically impossible—(*Math on Mutations*) Fortunately mutations are rare. They normally occur on an average of perhaps once in every ten million duplications of a DNA molecule.

Even assuming that all mutations were beneficial—in order for evolution to begin to occur in even a small way, it would be necessary to have, not just one, but a SERIES of closely related and interlocking mutations—all occurring at the same time in the same organism!

The odds of getting two mutations that are in some slight manner related to one another is the product of two separate mutations: ten million times ten million, or a hundred trillion. That is a 1 followed by 14 zeros (in scientific notation written as 1×10^{14}). What can two mutations accomplish? Perhaps a honeybee with a wavy edge on a bent wing. But he is still a honeybee; he has not changed from one species to another.

It is impossible for the mutations to lead to speciation, if one assumes the Darwinian model as the mechanism of speciation; however, the GMCMI model, in which the gross mutations might be involved with millions of DNA molecules simultaneously, makes the occurrence of speciation mathematically likely. For the details, see Chapter 5.

11 - time is no solution—But someone will say, "Well, it can be done—if given enough time." Evolutionists offer us 5 billion years for mutations to do the job of producing all the wonders of nature that you see about you. But 5 billion years is, in seconds, only 1 with 17 zeros (1×10^{17}) after it. And the whole universe only contains 1×10^{80} atomic particles. So there is no possible way that all the universe and all time past could produce such odds as would be needed for the task! Julian Huxley, the leading evolutionary spokesman of mid-twentieth century, said it would take 10^{3000} changes to produce just one horse by evolution. That is 1 with 3000 zeros after it! (Julian Huxley, *Evolution in Action*, p. 46).

Evolution requires millions of beneficial mutations all working closely together to produce delicate living systems full of fine-tuned structures, organs, hormones, and all the rest. And all those mutations would have to be non-random and intelligently planned! In no other way could they accomplish the needed task.

But, leaving the fairyland of evolutionary theory, to the real world, which only has rare, random, and harmful mutations, we must admit that mutations simply cannot do the job.

In the GMCMI model, speciation occurs instantly. The challenge is how to find them.

12 gene stability—It is the very rarity of mutations that guarantees the stability of the genes. Because of that, the fossils of ancient plants and animals are able to look like those living today.

"Mutations rarely occur. Most genes mutate only once in 100,000 generations or more." "Researchers estimate that a human gene may remain stable for 2,500,000 years."—*World Book Encyclopedia, 1966 Edition.*

Even 1% or less of mutation are stable and useful, that is enough to cause all species observed by assuming the GMCMI model is correct.

13 against all law—After spending years studying mutations, Michael Denton, an Australian research geneticist, finalized on the matter this way:

"If complex computer programs cannot be changed by random mechanisms, then surely the same must apply to the genetic programs of living organisms.

"The fact that systems [such as advanced computers], in every way analogous to living organisms, cannot undergo evolution by pure trial and error [by mutation and natural selection] and that their functional distribution invariably

167

conforms to an improbable discontinuum comes, in my opinion, very close to a formal disproof of the whole Darwinian paradigm of nature. By what strange capacity do living organisms defy the laws of chance which are apparently obeyed by all analogous complex systems?"—*Michael Denton, Evolution: A Theory in Crisis (1985), p. 342.*

It is true that the natural selection (speciation) is against all laws. The GMCMI model does not violate any natural laws.

14.**Syntropy**—This principle was mentioned in the chapter on natural selection; it belongs here also. Albert Szent-Gyorgyi is a brilliant Hungarian scientist who has won two Nobel Prizes (1937 and 1955) for his research. In 1977, he developed a theory which he called syntropy. Szent-Gyorgyi points out that it would be impossible for any organism to survive even for a moment, unless it was already complete with all of its functions and they were all working perfectly or nearly so. This principle rules out the possibility of evolution arising by the accidental effects of natural selection or the chance results of mutations. It is an important point.

"In postulating his theory of syntropy, Szent-Gyorgyi, perhaps unintentionally, brings forth one of the strongest arguments for Creationism—the fact that a body organ is useless until it is completely perfected. The hypothesized law of 'survival of the fittest' would generally select against any mutations until a large number of mutations have already occurred to produce a complete and functional structure; after which natural selection would then theoretically select for the organism with the completed organ."—*Jerry Bergman, "Albert Szent-Gyorgyi's Theory of Syntropy," in Up with Creation (1978), p. 337.*

I agree with Szent-Györgyi that it would be impossible for any organism to survive unless it is

168

completely perfected. The GMCMI model suggests all new species were completely functional at birth.

15.minor changes damage offspring the most—With painstaking care, geneticists have studied mutations for decades. An interesting feature of these accidents in the genes, called mutations, deals a stunning blow to the hopes of neo-Darwinists. Here, in brief, is the problem:

(1) Most mutations have very small effects; some have larger ones. (2) Small mutations cannot accomplish the needed task, for they cannot produce evolutionary changes. Only major mutational changes, with wide-ranging effects in an organism, can possibly hope to effect the needed changes from one species to another.

And now for the new discovery: (3) it is only the minor mutational changes, which harm one's descendants. The major ones kill the organism outright or rather quickly annihilate its offspring!

"One might think that mutants that cause only a minor impairment are unimportant, but this is not true for the following reason: A mutant that is very harmful usually causes early death or senility. Thus the mutant gene is quickly eliminated from the population. Since minor mutations can thus cause as much harm in the long run as a major ones, and occur much more frequently, it follows that most of the mutational damage in a population is due to the accumulation of minor changes."—*J.F. Crow, "Genetic Effects of Radiation," in Bulletin of the Atomic Scientists, January 1958, p. 20*

The GMCMI model proposes, instead of small mutations, that the gross mutations play an important role in speciation. Human infertility research has shown that persons with a gross mutation can have a normal life span, as explored in chapter 7.

16.would have to do it in one generation— Not even one major mutation, affecting a large number of organic factors, could accomplish the task of taking an organism across the species barrier. Hundreds of mutations—all positive ones,—and all working together would be needed to produce a new species. The reason: The formation of even one new species would have to be done all at once—in a single generation!

The GMCMI model suggests all mutations produce a new species in a single generation.

17.inconsequential accomplishments—A major problem here is that, on one hand, mutations are damaging and deadly; but on the other,—aside from the damage—they only directly change small features.

"Is it really certain, then, as the neo-Darwinists maintain, that the problem of evolution is a settled matter? I, personally, do not think so, and, along with a good many others, I must insist on raising some banal objections to the doctrine of neo-Darwinism "

.

Richard Goldschmidt was the geneticist who first proposed miraculous multimillion, beneficial mutations as the only possible cause of species crossover. (More on this later.) This is what he wrote about the inconsequential nature of individual mutations:

"Such an assumption [that little mutations here and there can gradually, over several generations, produce a new species] is violently opposed by the majority of geneticists, who claim that the facts found on the subspecific level must apply also to the higher categories. Incessant repetition of this unproved claim, glossing lightly over the difficulties, and the assumption of an

170

arrogant attitude toward those who are not so easily swayed by fashions in science, are considered to afford scientific proof of the doctrine. It is true that nobody thus far has produced a new species or genus, etc., by macromutation. It is equally true that nobody has produced even a species by the selection of micromutations."—*Richard Goldschmidt, in American Scientist (1952), p. 94.*

See question 15.

18.traits are totally interconnected—Experienced geneticists are well-aware of the fact that the traits contained within the genes are closely interlocked with one another. That which affects one trait will affect many others. They work together. Because of this, all the traits, in changed form, would have to all be there together—instantly,—in order for a new species to form!

"Each mutation occurring alone would be wiped out before it could be combined with the others. They are all interdependent. The doctrine that their coming together was due to a series of blind coincidences is an affront not only to common sense but to the basic principles of scientific explanation."—*A. Koestler, The Ghost in the Machine (1975), p. 129.*

See question 15 above.

19 too many related factors—There are far too many factors associated with each trait for a single mutation—or even several to accomplish the needed task. Mathematical probabilities render mutational species changes impossible of attainment.

"Based on probability factors, any viable DNA strand having over 84 nucleotides cannot be the result of haphazard mutations. At that stage, the probabilities are 1 in $4^{80} \times 10^{50}$.

The model proposes the gross mutations, not single mutations, are responsible for speciation. What we see are successful ones.

20 - reproductive changes low—Here is an extremely IMPORTANT point: Mutational changes in the reproductive cells occur far more infrequently than in the cells throughout the rest of the body. Only mutational changes within the male or female reproductive cells could affect oncoming generations.

"The mutation rates for somatic cells are very much higher than the rates for gametic cells. *"Biological Mechanisms Underlying the Aging Process," in Science, August 23, 1963, p. 694.*

The gross mutation occurs in the zygote stage. The mutation rates of zygotes might be higher than ones in somatic or gametic cells, and the mutation are also inheritable.

21.evolution requires increasing complexity—The theorists have decreed that evolution, by its very nature, must move upward into ever-increasing complexity, better structural organization, and completeness. Indeed, this is a cardinal dictum of evolutionists. Evolutionists maintain that evolution can only move upward toward more involved life-forms,—and that it can never move backward into previously evolved life-forms.

But, in reality, mutations, by their very nature, tear down, disorganize, crumble, confuse, and destroy.

According to the GMCMI model, mutations are always random and might create organisms with either upward or downward complexity. However, organisms with better structures might have a survival advantage over their counterparts, which make evolution seem to have the direction.

22 evolution requires new information—In order for a new organism to be formed by evolutionary change, new information banks must be emplaced. It is something like using a more advanced computer program; a "card" of more complicated procedural instructions must be put into the central processing unit of that computer. But the haphazard, random results of mutations could never provide this new, structured information.

"If evolution is to occur living things must be capable of acquiring new information, or alteration of their stored information."—*George Gaylord Simpson, "The Non-prevalence of Humanoids," in Science, 143, (1964), p. 772.*

New information can be generated in the process of the gross mutations.

23.evolution requires new organs—It is not enough for mutations to produce changes;—they must produce new organs! Billions of mutational factors would be required for the invention of one new organ of a new species, and this mutations cannot do.

"A fact that has been obvious for many years is that Mendelian mutations deal only with changes in existing characters . . No experiment has produced progeny that show entirely new functioning organs. And yet it is the appearance of new characters in organisms which mark the boundaries of the major steps in the evolutionary scale."—*H.G. Cannon, The Evolution of Living Things (1958).*

Just as the above.

24.evolution requires complicated networking—A relatively new field of scientific study is called "linkage," "linkage interconnections," or "networking." This is an attempt to analyze the network of interrelated factors in the body. I say, "an attempt," for there are millions of such linkages. Each structure or organ is related to another—and also to thousands of others._(A detailed study of this type of research will be found in Creation Research Society Quarterly, for March 1984, pp. 199-211. Ten diagrams and seven charts are included.)

Our concern here is that each mutation would damage a multi-link network. This is one of the reasons why mutations are always injurious to an organism.

The kidneys interconnect with the circulatory system, for they purify the blood. They also interconnect with the nervous system, the endocrine system, the digestive system, etc. But such are merely major systems. Far more is included. We are simply too fearfully and wonderfully made for random mutations to accomplish any good thing within our bodies.

The gross mutation can affect hundreds or even thousands genes simultaneously, and a complicated networking could be established in a single step.

25.visible and invisible mutations—"Visible mutations" are those genetic changes that are easily detectable, such as albinism, dwarfism, and hemophilia. Winchester explains: (1) For every visible mutation, there are 20 lethal ones, which are invisible! (2) Even more frequent than the lethal mutations would be the ones that damage but do not kill.

"Lethal mutations outnumber visibles by about 20 to 1. Mutations that have small harmful effects, the detrimental mutations, are even more frequent than the lethal ones."—A.M. Winchester, *Genetics*, 5th Edition (1977), p. 356.

Even a small percentage of "visible mutations" survive and become a noticeable species, which would be enough to generate all species we know.

26 never higher vitality than parent—Geneticists, who have spent a lifetime studying mutations, tell us that each mutation only weakens the organism. Never does the mutated offspring have more strength than the unmutated (or less mutated) parent.

"There is no single instance where it can be maintained that any of the mutants studied has a higher vitality than the mother species. It is, therefore, absolutely impossible to build a current evolution on mutations or on recombinations."—*N. Herbert Nilsson, Synthetische Artbildung (Synthetic Speciation) (1953), p. 1157 [italics his].*

Nature has many ways to generate mutants. That geneticists never observe strong mutant offspring does not mean they do not exist. Laboratory conditions might be more harmful than one in the natural condition.

27 mutations are not producing species change—Theory, theory, lots of theory, but it just isn't happening!

"No matter how numerous they may be, mutations do not produce any kind of evolution."—*Pierre Paul Grasse, Evolution of Living Organisms (1977), p. 88.*

"It is true that nobody thus far has produced a new species or genus, etc., by macromutation [a combination of many mutations]; it is equally true that nobody has produced even a species by the selection of micromutation [one or only a few mutations]."—*Richard B. Goldschmdt, "Evolution, As Viewed by One Geneticist, "American Scientist, January 1952, p. 94.*

A "nascent organ" is one that is just coming into existence. None have ever been observed.

The non-gross mutations would never produce a species change under Darwin's theory. Only gross mutations would lead to new species.

28.gene uniqueness forbids species change— The very fact that each species is so different than the others—forbids the possibility that random mutations could change them into new species. There are million of factors which make each species different than all the others. The DNA code barrier that would have to be crossed is simply too immense.

"If life really depends on each gene being as unique as it appears to be, then it is too unique to come into being by chance mutations."—*Frank B. Salisbury, "Natural Selection and the Complexity of the Gene," Nature, October 25, 1969, p. 342.*

Just as the above.

Appendix II. Conventions

In this book, the following assumptions are made.

1. This book only discusses speciation in animals.

2. Natural selection is interchangeable with the modern synthesis, Darwinism or new-Darwinism.

3. Macroevolution is speciation at the species level and above; whereas microevolution is the evolution within species.

4. The common descent and microevolution are assumed true, even there might be more than one common descent.

Appendix III. Abbreviations

BCS	biological concept of species
DNA	deoxyribonucleic acid
mtDNA	mitochondrial DNA
EP	punctuated equilibrium
GMCMI	gross mutations in cluster and mutants inbreeding
ID	intelligent design
MHC	major histocompatability complex
MIST	mixed identical supertwins
MISTWGM	mixed identical supertwins with gross mutation
MMIZ	multiple mixed identical zygotes
MMIZWGM	multiple mixed identical zygotes with gross mutation
PSN	pre-species novelties
RNA	ribonucleic acid
mRNA	message RNA
rRNA	ribosomal RNA
tRNA	transport RNA
SMRS	specific mate recognition system
SARS	severe acute respiratory syndrome

Bibliography

Chapter 1

Kalantari, P., Sepehri, H., Behjati, F., Ashtiani, Z. O. and Akbari, M. T. (2003). "Chromosomal studies of male infertility." Genetika 39 (3): p. 423-6.

Murray, P. G. a. Y., Lawrence S. (2001). "Epstein-Barr virus infection: basis of malignancy and potential for therapy." Exp. Rev. Mol. Med. (November): p. 15.

Ray, F. A. (1995). "Simian Virus 40 Large T Antigen Induces Chromosome Damage That Precedes and Coincides with Complete Neoplastic Transformation". Oncogenic Mechanisms of DNA Tumor Viruses. G. Barbanti-Brodano, Bendinelli, M. and Friedman, H. New York, Plenum Publishing Corp. p. 15-26.

Sankoff, D., Deneault, M., Turbis, P. and Allen, C. (2002). "Chromosomal distributions of breakpoints in cancer, infertility, and evolution." Theor Popul Biol 61 (4): p. 497-501.

Chapter 3

Campbell (1996). "Biology".4th eds. Menlo Park, The Benjuamin/Cummings Publishing Company, Inc. p. 444.

Campbell, N. R., Jane (2002). "Biology".6th eds. Menlo Park, The Benjuamin/Cummings Publishing Company, Inc. p. 468-475.

Futuyma, D. J. (1986). "Evolutionary biology".2nd. Sunderland, Mass., Sinauer Associates. p. 12.

Hall, B. K. (1995). "Atavisms and atavistic mutations." Nat Genet 110 (22): p. 1126-27.

Mallet, J. (1995). "A species definition for the Modern Synthesis." Trends Ecol. Evol. (10): p. 294-299.

Mayr, E. (2001). "What evolution is". New York, Basic Books. p. 180.

Paterson, H., Ed. (1986). "Environment and Species". Baltimore, MD, John Hopkins University Press.

Wells, J. (2000). "Icons of evolution: science or myth? why much of what we teach about evolution is wrong". Washington, DC, Regnery Publishing, Inc. p. 219-21.

Chapter 4

Andeson, N. (1970). "Evolutionary Significance of Virus Infection." Nature 227: p. 1347.

Campbell (1996). "Biology". Menlo Park, The Benjuamin/Cummings Publishing Company, Inc. p. 398.

Clark, A. H. (1930). "The New Evolution: Zoogenesis". p. 100, 114, 189, 196, 211.

Darwin, C. (1860). "On the origin of species by means of natural selection".2th. New York, D. Appleton and Company. p. 178.

Darwin, C. and Darwin, F. (1887). "The life and letters of Charles Darwin". New York, D. Appleton and company. p. 229.

Dobzhansky, T. (1973). "Nothing in biology makes sense except in the light of evolution." American Biology Teacher (35): p. 125-9.

Futuyma, D. J. (1986). "Evolutionary biology".2nd. Sunderland, Mass., Sinauer Associates. p. 12.

Goldschmidt, R. B. (1940). "*The Material Basis of Evolution*". New Haven, Yale University Press.

Jaxon, E. (2002). "Perspectives on Evolution." http://members.soltec.net/~jaxon/Writing/Evolution.html.

Johnson, P. E. (1997). "Defeating Darwinism by opening minds". Downers Grove, InterVarsity Press. p. 58-59.

Kuhn, T. S. (1962). "The structure of scientific revolutions". Chicago, University of Chicago Press.

Lewin, R. (1980). "Evolutionary theory under fire." Science 210 (4472): p. 883-87.

Mayr, E. (1991a). "One Long Argument, *Charles Darwin and the Genesis of Modern Evolutionary Thoughts*". Cambridge, Massachusetts, Harvard University Press. p. 36-37.

Mayr, E. (1991b). "One Long Argument, *Charles Darwin and the Genesis of Modern Evolutionary Thoughts*". Cambridge, Massachusetts, Harvard University Press. p. 35.

Milner, R. (1990). "The Encyclopedia of Evolution, *Humanity's Search for Its Origins*". New York, Henry Holt and Company. p. 396.

Paley, W. (1802). "Natural theology". London, Printed for R. Faulder by Wilks and Taylor.

Rostand, J. (1961). "The Orion book of evolution". New York, Orion Press. p. 64.

Ruse, M. (1981). "Darwin's Theory: An Exercise in Science". New Scientist: p. 828

Simpson, G. G. (1951). "The meaning of evolution".A special rev. and abridged. New York, New American Library. p. 134-135.

Taylor, G. R. (1983). "The great evolution mystery".1st U.S. New York, Harper & Row. p. 55.

Chapter 5

Cain, A. J. (1954). "Animal Species and Their Evolution". London, Hutchinson University Library.

Davison, J. A. (1984). *"Semi-meiosis as an evolutionary mechanism."* J. Theor. Biol. 111 (3): p. 725-35.

Chickens First

Davison, J. A. (2000). "An Evolutionary Manifesto: A New Hypothesis for Organic Change." Jun 15. http://www.uvm.edu/-jdavison/.

Digby, L. (1912). "The cytology of Primula kewensis and of other related Primula hybrids." Ann. Bot. 26: p. 357-388.

Goldschmidt, R. B. (1940). "*The Material Basis of Evolution*". New Haven, Yale University Press.

Irion, R. (1996). "Humble Pocket Gophers Shed Light on the Genetic Fortitude of Cheetahs". Genetics

Johannsen, W. (1911). "The Genotype Conception of Heredity." American Naturalist (45): p. 129-159.

Mayr, E. (2001). "What evolution is". New York, Basic Books. p. 79.

Newton, W. C. F. a. C. P. (1929). "Primula kewensis and its derivatives." Genetics 20: p. 405-467.

O'Brien, S. J. and Yuhki, N. (1999). "Comparative genome organization of the major histocompatibility complex: lessons from the Felidae." Immunol Rev. 167: p. 133-44.

Ridley, M., Ed. (1996). "Evolution". Cambridge, Mass., USA, Blackwell Science. p. 418.

Schindewolf, O. H. and Reif, W.-E. (1993). "Basic questions in paleontology : geologic time, organic evolution, and biological systematics". Chicago, University of Chicago Press.

Stanley, S. M. (1979). "Macroevolution: Pattern and Process". p. 159.

Chapter 6

Durand, J., Keller, N., Renard, G., Thorn, R. and Pouliquen, Y. (1993). "Residual cornea and the degenerate eye of the cryptophthalmic Typhlotriton spelaeus." Cornea 12 (5): p. 437-47.

Hunt, K. (1995). "Horse Evolution." January 4, 1995. http://www.talkorigins.org/faqs/horses/horse_evol.html

Jeffery, W. R. (2001). "Cavefish as a model system in evolutionary developmental biology." Dev Biol. 231 (1): p. 1-12.

Kos, M., Bulog, B., Szel, A. and Rohlich, P. (2001). "Immunocytochemical demonstration of visual pigments in the degenerate retinal and pineal photoreceptors of the blind cave salamander (Proteus anguinus)." Cell Tissue Res. 303 (1): p. 15-25.

Mayr, E. (1942). "Systematics and the origin of species from the viewpoint of a zoologist". New York, Columbia University Press. .

Mayr, E. (2001). "What evolution is". New York, Basic Books. p. 190.

Nishikimi, M., Fukuyama, R., Minoshima, S., Shimizu, N. and Yagi, K. (1994). "Cloning and chromosomal mapping of the human nonfunctional gene for L-gulono-gamma-lactone oxidase, the enzyme for L-ascorbic acid biosynthesis missing in man." J. Biol. Chem. 269 (18): p. 13685-88.

Nishikimi, M., Kawai, T. and Yagi, K. (1992). "Guinea pigs possess a highly mutated gene for L-gulono-gamma-lactone oxidase, the key enzyme for L-ascorbic acid biosynthesis missing in this species." J. Biol. Chem. 267 (30): p. 21967-72.

Ohta, Y. and Nishikimi, M. (1999). "Random nucleotide substitutions in primate nonfunctional gene for L-gulono-gamma-lactone oxidase, the missing enzyme in L-ascorbic acid biosynthesis." Biochim Biophys Acta 1472 (1-2): p. 408-11.

Paley, W. (1802). "Natural theology". London, Printed for R. Faulder by Wilks and Taylor.

Sheldon, P. R. (1987). "Parallel gradualistic evolution of Ordovician trilobites." Nature 330 (6148): p. 561-63.

Wells, J. (2000a). "Icons of evolution: science or myth? : why much of what we teach about evolution is wrong". Washington, DC, Regnery Publishing, Inc.

Wells, J. (2000b). "Icons of evolution: science or myth? : why much of what we teach about evolution is wrong". Washington, DC, Regnery Publishing, Inc. p. 181.

Chapter 7

Crick, F. (1981). "*Life Itself: Its Origin and Nature*". New York, Simon and Schuster.

Darwin, C. (1871). "The descent of man, and selection in relation to sex". New York, D. Appleton and company.

Mayr, E. (1991). "One Long Argument, *Charles Darwin and the Genesis of Modern Evolutionary Thoughts*". Cambridge, Massachusetts, Harvard University Press. p. 159-162.

Mayr, E. (2001). "What evolution is". New York, Basic Books. p. 272.

Shklovskii, I. S. and Sagan, C. (1966). "Intelligent life in the universe". San Francisco, Holden-Day.

Westoll, T. S., Andrews, S. M., Miles, R. S. and Walker, A. D. (1977). "Problems in vertebrate evolution : essays presented to Professor T. S. Westoll". London ; New York, published for the Linnean Society of London by Academic Press.

Chapter 8

"Evolution, Morality, and Violence." http://evolution-fact.org/c19a.htm.

Bowler, P. J. (1989). "Evolution : the history of an idea".Rev. Berkeley, University of California Press. p. 342-43.

Dawkins, R. (1976). "The Selfish Gene". New York, Oxford University Press. p. 7-8.

Eldredge, N. (2000). "The triumph of evolution and the failure of creationism". New York, W.H. Freeman. p. 154.

Greene, J. C. (1981). "Science, Ideology, and World View: Essays in the History of Evolutionary Ideas". Berkeley, Los Angeles, London, University of California Press.

Hamilton, W. D. (1964). "The Genetical Evolution and Social Behavior, I and II." J. Theor. Biol. (7): p. 1-32.

Kropotkin, P. A. and Huxley, T. H. (1972). "Mutual aid, a factor of evolution". New York, Garland Publisher, Inc.

Mayr, E. (2001a). "What evolution is". New York, Basic Books. p. 259.

Mayr, E. (2001b). "What evolution is". New York, Basic Books. p. 128-30.

Newman, V. R., Ed. (1995). "Evolution and Genetics". p. 272.

Ostrander, G. M. (1971). "The Evolutionary Outlook: 1875-1900". p. 5.

Romanes, G. (1878). "A Candid Examinatiion of Theism".

Rose, M. (1998a). "Darwin's Spectre, *Evolutionary Biology in the Modern World*". New Jersey, Princeton University Press. p. 98-99.

Rose, M. (1998b). "Darwin's Spectre, *Evolutionary Biology in the Modern World*". New Jersey, Princeton University Press. p. 142.

Ruse, M. (1982). "Darwinism defended : a guide to the evolution controversies". Reading, Mass., Addison-Wesley, Advanced Book Program/World Science Division. p. 356.

Chapter 9

Broom, R. (1951). "Finding the missing link". London, Watt. p. 107.

Cann, R. L., Stoneking, M. and Wilson, A. C. (1987). "Mitochondrial DNA and human evolution." Nature 325 (6099): 31-6.

Huxley, J. (1963). *Evolution, The Modern Synthesis*". London, Allen and Unwin. p. 571.

Kuhn, T. S. (1962). "The structure of scientific revolutions". Chicago, University of Chicago Press.

Popper, K. (1968). "Conjectures and refutations: the growth of scientific knowledge". New York, Harper & Row.

Zuckerkandl, E. and Pauling, L. (1965). "Molecules as documents of evolutionary history." J. Theor. Biol. 8 (2): p. 357-66.

Chapter 10

Bateson, W. (1913). "Problems of genetics". New Haven, Yale University Press.

Berg, L. (1969). "Nomogenesis; or, Evolution Determined by Law." Cambridge, M.I.T. Press.

Dobzhansky, T. (1973). "Nothing in biology makes sense except in the light of evolution." American Biology Teacher (35): p. 125-9.

Chickens First

Grassé, P. (1977). *"Evolution of Living Organisms: Evidence for a New Theory of Transformation"*. New York, Academic Press. p. 119.

Milner, R. (1990). "The Encyclopedia of Evolution, *Humanity's Search for Its Origins"*. New York, Henry Holt and Company. p. 336.

Monod, J. L. (1997). "On the molecular theory of evolution". Evolution. M. Ridley. Oxford, New York, Oxford University Press p. 390.

Punnett, R. (1915). "Mimicry in butterflies". Cambridge, University Press.

Closing Marks

Gold, P.(2000). The Washington Times, End of Darwinism? October 25.

Index